Animal cells are a relative newcomer to the field of biotechnology, but they are increasing in importance with the development of genetically engineered cell lines. This book is a guide to the culture requirements, preservation and the international organisations from which help and information can be obtained. It will be indispensible to biologists whether in industry or in academia. It will allow both existing scientists in the field and others entering the discipline for the first time to understand some of the basic principles underlying animal cell technology and all those other, seemingly unrelated, but entirely essential pieces of information necessary to work efficiently and effectively. The authors are all practicing scientists with considerable expertise. They have drawn upon this to make each contribution a mine of information which is impossible to obtain elsewhere.

An international initiative by the
World Federation for Culture Collections,
with financial support from UNESCO

LIVING RESOURCES FOR BIOTECHNOLOGY

Animal cells

Titles in the series
Animal cells
Bacteria
Filamentous fungi
Yeasts

LIVING RESOURCES FOR BIOTECHNOLOGY

Animal Cells

Edited by

A. Doyle, R. Hay and B. E. Kirsop

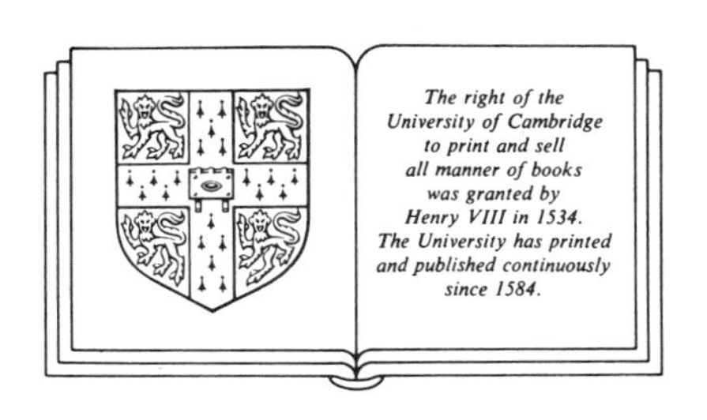

CAMBRIDGE UNIVERSITY PRESS
Cambridge
New York   Port Chester   Melbourne   Sydney

Published by the Press Syndicate of the University of Cambridge
The Pitt Building, Trumpington Street, Cambridge CB2 1RP
40 West 20th Street, New York, NY 10011, USA
10 Stamford Road, Oakleigh, Melbourne 3166, Australia

First published 1990

Printed in Great Britain at the University Press, Cambridge

British Library cataloguing in publication data
Animal cells.
1. Biotechnology. Applications of animal cultured cells
I. Doyle, A. (Alan) II. Hay, R. III. Kirsop, B. E.
660.6

Library of Congress cataloguing in publication data
Animal cells/edited by A. Doyle, R. Hay, and B. E. Kirsop.
p. cm. – (Living resources for biotechnology)
Includes bibliographical references.
Includes index.
1. Animal cell biotechnology. I. Doyle, A. (Alan) II. Hay, Robert. III. Kirsop, B. E. IV. Series.
[DNLM: 1. Biotechnology. 2. Cells, Cultured. 3. Tissue Culture. QH 585 A5984]
TP248.27.A53A55 1990
660'.6–dc20 90-2025 CIP

ISBN 0 521 35223 1 hardback

CONTENTS

CONTRIBUTORS

Allner, K. Quality Control Group, Public Health Laboratory Service Centre for Applied Microbiology & Research, Porton Down, Salisbury, Wiltshire SP4 0JG, UK (Chapter 3)

Allsopp, D. CAB International Mycological Institute, Ferry Lane, Kew, Surrey TW6 3AF, UK (Chapter 7)

Blaine, L. D. Hybridoma Data Bank, 12301 Parklawn Drive, Rockville, Maryland 20852, USA (Chapter 2)

Bousfield, I. J. National Collections of Industrial and Marine Bacteria Ltd, 23 St Machar Drive, Aberdeen AB2 1RY, UK (Chapter 6)

DaSilva, E. J. United Nations Educational Scientific and Cultural Organisation, Division of Scientific Research and Higher Education, 7 Place de Fontenoy, 75700 Paris, France (Chapter 8)

Doyle, A. European Collection of Animal Cell Cultures, Public Health Laboratory Service Centre for Applied Microbiology & Research, Porton Down, Salisbury, Wiltshire SP4 0JG, UK (Chapters 1, 3, 4, 5)

Griffiths, J. B. Animal Cell Technology Group, Public Health Laboratory Service Centre for Applied Microbiology & Research, Porton Down, Salisbury, Wiltshire SP4 0JG, UK (Chapter 4)

Hay, R. American Type Culture Collection, 12301 Parklawn Drive, Rockville, Maryland 20852, USA (Chapter 1)

Kirsop, B. E. Microbial Strain Data Network, Institute of Biotechnology, Cambridge University, 307 Huntingdon Road, Cambridge CB3 0JX, UK (Chapter 8)

Krichevsky, M. I. Microbial Systematics Division, Epidemiology and Oral Disease Prevention Program, National Institute of Dental Research, Bethesda, Maryland 20892, USA (Chapter 2)

Morris, C. B. European Collection of Animal Cell Cultures, Public Health Laboratory Service Centre for Applied Microbiology & Research, Porton Down, Salisbury, Wiltshire, SP4 0JG, UK (Chapters 4, 5)

Mowles, J. M. European Collection of Animal Cell Cultures, Public Health Laboratory Service Centre for Applied Microbiology & Research, Porton Down, Salisbury, Wiltshire, SP4 0JG, UK (Chapter 5)

Newell, D. G. Pathology Division, Public Health Laboratory Service Centre for Applied Microbiology & Research, Porton Down, Salisbury, Wiltshire SP4 0JG, UK (Chapter 4)

Ohno, T. Riken Gene Bank, RIKEN (The Institute of Physical and Chemical Research), 3-1-1 Yatabe, Koyadai, Tsukuba Science City 305, Japan (Chapter 1)

Simione, F. P. American Type Culture Collection, 12301 Parklawn Drive, Rockville, Maryland 20852, USA (Chapter 7)

Sugawara, H. Life Science Research Information Section, RIKEN, Wako, Saitama 351-01, Japan (Chapters 1, 2)

SERIES INTRODUCTION

The rapid advances taking place in biotechnology have introduced large numbers of scientists and engineers to the need for handling microorganisms, often for the first time. Questions are frequently raised concerning sources of cultures, location of strains with particular properties, requirements for handling the cultures, preservation and identification methods, regulations for shipping, or the deposit of strains for patent purposes. For those in industry, research institutes or universities with little experience in these areas, resolving such difficulties may seem overwhelming. The purpose of the World Federation for Culture Collections' (WFCC) series, Living Resources for Biotechnology, is to provide answers to these questions.

Living Resources for Biotechnology is a series of practical books that provide primary data and guides to sources for further information on matters relating to the location and use of different kinds of biological material of interest to biotechnologists. A deliberate decision was taken to produce separate volumes for each group of microorganism rather than a combined compendium, since our enquiries suggested that inexpensive specialised books would be of more general value than a larger volume containing information irrelevant to workers with interests in one particular type of organism. As a result each volume contains specialised information together with material on general matters (information centres, patents, consumer services, the international coordination of culture collection activities) that is common to each.

The WFCC is an international organisation concerned with the establishment of microbial resource centres and the promotion of their activities. In addition to its primary role of coordinating the work of culture collections through the world, the committees of the WFCC are active in a number of areas of particular relevance to biotechnology, such as patents, microbial information centres, postal and quarantine regulations, educational and conservation matters (see Chapter 8). The Education Committee of the WFCC proposed the preparation of the current volumes.

The WFCC is concerned that this series of books is of value to biotechnologists internationally, and the authors have been drawn from specialists throughout the world. The close collaboration that exists between culture collections in every continent has made the compilation of material for the books a simple and pleasurable process, since the authors and contributors are for the most part colleagues. The Federation hopes that the result of their labours has produced valuable source books that will not only accelerate the progress of biotechnology, but will also increase communication between culture collections and their users to the benefit of both.

Barbara Kirsop
President, World Federation
for Culture Collections

PREFACE

Until recently, the exploitation of animal cell cultures in biotechnological manufacturing has lagged behind the use of microorganisms such as bacteria, filamentous fungi and yeasts. One reason for this is the fastidious growth requirements of animal cells, which require complex and expensive growth media and elaborate culture systems. Further reasons for this slow progress are the slow growth characteristics of animal cells with doubling times of 24 hours (compared with as little as 30 minutes for some bacteria) and the generally low productivity levels.

Before the development of hybridomas (producing monoclonal antibodies) and genetically manipulated cell lines, the major industrial use for cell lines was as a substrate for virus production in viral vaccine manufacture (e.g. poliomyelitis, measles and rubella). New developments have radically altered the picture. The potential for monoclonal antibodies in diagnosis and therapy, together with biochemical manufacturing purification processes, is enormous and has become the basis for new industrial enterprises.

Because of the difficulty of obtaining secretion of some mammalian proteins in an active form from genetically engineered bacteria, another developing area is the engineering of animal cells for the production of therapeutic products. There is also growth in the use of cell lines for the production of immunoregulatory proteins, including interferon and interleukins, and the search is still continuing for suitable cell substrates. Interest in animal cells as tools for biotechnology has, therefore, never been greater.

The aim of this volume is to bring together as much useful information as possible for those who may be newcomers to the field of animal cell technology. The result is an international collaborative effort, with the strengths and weaknesses associated with such initiatives. The speed of developments in this fast moving field inevitably leads to source books becoming out of date before publication; the authors therefore apologise for any omissions which will be rectified in future editions.

Alan Doyle
European Collection of Animal Cell Cultures

ACKNOWLEDGEMENTS

Particular thanks are due to the following who provided useful help and information:
Dr G. Dunbar, Cell Repositories, Montreal Children's Hospital, Canada; Mr J. Gibson, Central Public Health Laboratory, Colindale, London; Dr A. Greene, NIA, NIGMS Coriell Institute for Medical Research, Camden, New Jersey, USA; Dr M. Mathieu, Human Genetic Cell Repository, Hospital Debrousse, Lyon, France; Dr Mizusawa, JCRB Cell Bank, National Institute of Hygienic Sciences, Kami-Yoga, Setagaya-ku, Tokyo, Japan; Mrs Jean Stanley, Commonwealth Serum Laboratories, Parkville, Victoria, Australia; Dr Masao Takeuchi, Curator of Animal Cell Lines, Institute for Fermentation, Osaka, Japan.

I would also like to pay a special tribute to one of our contributors, Mr Keith Allner, a friend and colleague at PHLS CAMR, who sadly died before publication of this volume. He contributed a wealth of experience to works such as this and he will be greatly missed.

Alan Doyle

1

Resource Centres

A. DOYLE, R. HAY, T. OHNO and H. SUGAWARA

1.1 Introduction

Public cell culture collections came about in response to a widespread need for well characterised, microbe-free seed stocks derived from cultures supplied by cell line originators. The explosion of research on viruses in the mid to late 1950s, enabled by the extensive use of cell culture, led to exploitation of the techniques, often without awareness of the need for critical, cell line quality control. By 1960 the problems of cellular and microbial contamination of cell lines had become so acute that scientists in the United States banded together to establish a bank of tested cells. Soon they realised that the application of sensitive and extensive quality control tests would be desirable. These efforts coupled with improving preservation technology were carried out internationally (Stevenson, 1963).

Over the past 25 years millions of dollars have been invested in the cell banking programmes over and above the research costs of the developed cell lines. These funds have been willingly provided by granting agencies in the expectation of insuring the investment in research that uses cells as model systems.

While the rationale for development and use of well organised collections is understood by many laboratory scientists, poorly characterised cell stocks for use in research studies are still exchanged all too frequently. Thus it is important to review periodically the potential pitfalls associated with the use of microbial and cell stocks obtained and processed casually, to increase and reinforce awareness of the problem within the scientific community.

Numerous instances of the exchange of cell lines contaminated with cells of other species have been documented, and published by others

(Nelson-Rees & Flandermeyer, 1977; Harris *et al.*, 1981; Nelson-Rees *et al.*, 1981a, b). Similarly, the problem of intraspecies cross-contamination among cultured human-cell lines has been recognised for over 15 years and detailed reviews are available on the subject (Gartler, 1968; Nelson-Rees *et al.*, 1974, 1981a, b; Lavappa, 1978). The loss of time and research funds as a result of these problems is incalculable.

Although bacterial and fungal contaminations represent an added concern, in most instances they are overt and easily detected and are therefore of less serious consequence than the more insidious contamination by mycoplasma. That the presence of these microorganisms in cultured cell lines often completely negates research findings has been emphasised over the years by Barile *et al.* (1973), Hopps *et al.* (1973) and McGarrity (1982). Still, the difficulties of detection and prevalence of contaminated cultures in the research community suggest that repeated restatements are warranted.

The methodology for characterisation of cell lines varies somewhat among banking agencies. For example, precise details on procedures used at the American Type Culture Collection (ATCC) for the acquisition, preservation, characterisation, cataloguing and distribution of cell lines have been widely publicised. They are available elsewhere and need only be cited at this point (Scherer, 1962; Stulberg *et al.*, 1970; Hay *et al.*, 1982; Hay, 1984a, 1985, 1986). In addition, techniques used by other institutions with like goals will be described in the forthcoming chapters of this handbook. Therefore, only a general scheme to outline the steps recommended for addition of a new cell line will be presented and discussed (Figure 1.1).

Most investigators submit starter cultures without charge and with minimal restrictions on distribution. For some component lines, recipients must sign a release form indicating that the cultures will not be redistributed or otherwise used for commercial purposes.

Required data for general submission of lines include a statement of exact species, strains, tissue sources, and precise description of original and subsequent dissociation procedures. Information on the growth medium and supplements should be included. A statement of specific functional characteristics, which would make appropriate strains uniquely valuable, also is required. In addition, preference is indicated for supply of detailed histories on each strain; information concerning date of establishment, sex, race, and age of donor; isolation procedures; inoculation densities; number of passages, and an indication of the average number of cell generations accrued. Information on karyology and survival potential of the cell lines is also solicited.

After review of the credentials of each given line and consultation with advisors, decisions are made to proceed with initial characterisations. Generally, starter cultures or ampoules are obtained from the donor, and progeny are propagated according to instructions to yield the first 'token' freeze. Cultures derived from such token material are then subjected to the minimal, but critical, characterisations. These include a series of tests for microbial contamination, including mycoplasma, plus fluorescent antibody staining and/or isoenzyme analysis to verify species (see Chapter 5).

If these tests suggest that the cell line is acceptable, it is expanded to produce seed and distribution stocks. Antibiotics are not used in the culture media at this stage so as not to mask infection. Note especially that the major characterisation efforts are applied to cell populations in the initial seed stock of ampoules. The distribution stock consists of ampoules that are distributed on request; reference seed stock, on the other hand, is retained to generate further distribution stocks as the initial stock becomes depleted.

Fig. 1.1. Scheme illustrating accessioning procedures used at the ATCC. Note that some or all of the characterisations listed on the *left-hand side* are performed depending on the category to which the line is assigned.

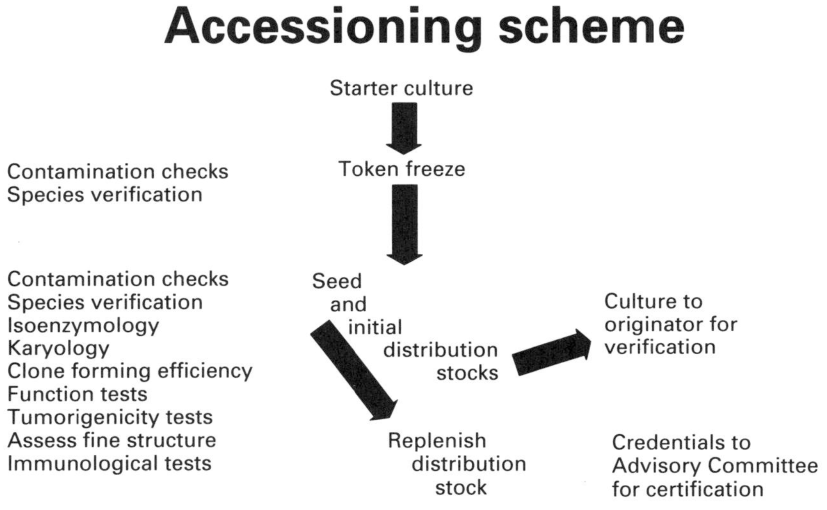

Although this procedure was designed to suit the needs of a large central repository, it is also applicable to smaller laboratories. Even where the number of cell lines and users may be limited, it is important to separate 'seed stock' from 'working or distribution stock'. If this is not done the frequent replacement of cultured material, recommended to prevent phenotypic drift or senescence, may deplete valuable seed stock which may be difficult and expensive to replace or, indeed, may be no longer available.

It is important to recognise that the characterised seed stock serves as a frozen 'reservoir' for production of distribution stocks over the years. Because seed stock ampoules are used to generate new distribution material, recipients can be assured that the cultures obtained closely resemble those received 2, 5, 10 or even more years previously. This is a most critical consideration for design of cell banking procedures. The seed stock is the most valuable of collection material, and records on all of the characterisations performed should relate directly back to the immediate progeny from these stocks.

The extent and types of characterisations performed vary depending upon the cell line, the category under which it is banked and the purpose for which the stock is intended. A description of the various components within the ATCC cell line Repository has been published elsewhere (Hay, 1984b). It is strongly recommended that a laboratory distribute, or accept for use from any source, only those cell lines that have both been thoroughly tested for microbial contaminants and had their species of origin verified.

Ideally the karyology of the cell line should also be determined and compared with that of other lines having similar characteristics. This latter step can significantly decrease the likelihood of intraspecies cellular cross-contamination (Nelson-Rees *et al.*, 1981b).

Cell banks rely on advice from *ad hoc* consultants and more formally organised committees. The degree and mode of interaction vary somewhat depending upon the department, disciplines, and source of support for differing accessioning programmes. In certification of cell lines, for example, input is requested concerning lines to be added, characterisations to be applied, and descriptions of the cellular material eventually banked. Specialists are often consulted at points during the accessioning procedures and especially for the final certification process. In this case, complete descriptions and pertinent data are often circulated for critical review. The cell lines and specific descriptions are considered acceptable if a majority of the experts indicate concurrence.

A policy of returning both a typical distribution ampoule or live culture plus the proposed catalogue description and data description on each cell line to its respective donor has been adopted by major banks. The donating scientist is thus given the opportunity to examine material prepared at the bank to verify that the essential characteristics are retained. Descriptions on cell cultures are also returned to the donor for suggested additions or revisions.

Above are summarised the key considerations for producing and certification of cell line stocks to be used in research or distributed for scientific study. The chapters which follow give additional information not only on the many international sources for cells but also on the characterisation of the material, culture maintenance and scale-up, safety in culture handling, information data banks, patenting of cell lines and international coordinating organisations.

1.2 Resource centres

With the increased biotechnological use of animal cells, there is a parallel increase in the number of culture collections and other resource centres. Some are listed in the World Data Center's *Directory of Collections of Cultures of Microorganisms* (V. F. McGowan & V. B. D. Skerman, 1986) available from the United Nations Environment Programme, Information Service, PO Box 30553, Nairobi, Kenya, or from the World Data Center (WDC), see Chapter 2. Other information can be sought from the data bases now being developed, for example the Committee on Data for Science and Technology's (CODATA) Hybridoma Data Bank (see Chapter 2).

Animal cell technology is now used in a multitude of disciplines in the biological sciences. As with culture collections of microorganisms, the activities of animal cell banks stretch across medical, veterinary, agricultural and biotechnological interests. As fields related to this technology are constantly developing, it is increasingly important to ensure that the information available is as comprehensive as possible and the use of electronic systems is essential for the effective distribution of data (see Chapter 2).

As well as the traditional culture collections exemplified by the ATCC, the needs of animal cell technologists have led to the development in many institutes of resource centres built around specific areas of interest, such as human genetics. Increasingly, funds are becoming available to establish resource centres and new developments are occurring throughout the world.

Each collection strives to publish the most up-to-date information on the cell lines it maintains. Centres may be contacted for current catalogues or enquiries may be made regarding particular cell lines. In addition to the supply of cultures, which may be supplied either in the form of growing cultures or frozen stocks, collections generally offer other unique services. These may include the characterisation and quality control of cell cultures. Many centres also run workshops and training courses in areas of specific interest to biotechnologists and may undertake contract research.

Collections may be classed either as general collections or specialist collections, depending on the activities they carry out.

1.3 General collections (listed alphabetically by geographical region)

Australia

Commonwealth Serum Laboratories
45 Poplar Road
Parkville
Victoria 3052
Australia
Telephone: 389-1911
Telex: AA 32789
Current holdings: 41 cell lines

Commonwealth Serum Laboratories was established in 1916 to produce and distribute essential vaccines and other important biological products. One of its activities is the maintenance and supply of authenticated cultures originating in Australia and overseas. The collection was established in 1958.

Bulgaria

The National Bank for Industrial Micro-organisms and Cell Cultures (NBIMCC)
Blvd. 'Lenin' 125 BL 2
V Floor
Sofia
Bulgaria
Telephone: 7208-65

The collection is recognised as an International Depository under the Budapest Treaty for the deposit of cell lines for patent purposes.

Canada

The Repository for Human Cell Strains and the Cell Repository for Neuromuscular Diseases
The McGill University
Montreal Children's Hospital Research Institute
2300 rue Tupper Street
Montreal
Quebec H3H 1P3
Canada
Telephone: (514) 934-4400 (ext. 2549)

The Repository for Human Cell Strains was established in 1971, with the Neuromuscular Disease Repository added in 1983. The collections comprise a bank of human skin fibroblasts, amniotic cell and foetal cell cultures from normal subjects and patients with chromosomal and biochemical abnormalities.

European Collection for Biomedical Research

A new grouping has been formed which brings together cell banks with a joint repository of Human cell lines derived for genetic studies. The work of these cell banks is coordinated under the Commission of the European Community's Biotechnology Action Programme. A joint catalogue is planned as a part of the project. The group includes:

Department of Cell Biology and Genetics
Erasmus University
PO Box 1783
NL-3000 DR Rotterdam
The Netherlands

Institut für Immunogenetik
Universitätsklinikum Essen
Virchowstrasse 171
D-4300 Essen 1
Federal Republic of Germany

Instituto Nazional per la Recerca sul Cancro
Viale Benedetto XV, 10
I-16132 Genova
Italy

European Human Cell Bank
ECACC
PHILS Centre for Applied Microbiology & Research
Porton Down
Salisbury SP4 0JG
UK

France

Collection Nationale de Cultures de Microorganismes
Institut Pasteur
25 rue du Dr Roux
F-75724 Paris Cédex 15
France
Telephone: (1) 45-68-82-51
Telex: 2500609 PASTEUR F

The Culture Collection at the Institut Pasteur is recognised as a depository for patent purposes for animal cells and viruses by the European Patent Office.

The Human Genetic Cell Repository
Hospices Civils de Lyon
Hospital Debrousse
29 rue Soeur Bourier
F-69322 Lyon Cédex 05
France
Telephone: (76) 78-25-16-65
Current holdings: over 1000

The Human Mutant Cell Repository has been established since 1972 in the cell culture department of the Biochemistry Laboratory in Debrousse Children's Hospital. The collection is comprised of human skin fibroblasts together with trophoblasts, amniotic fluid cultures and other foetal-derived cells.

Federal Republic of Germany

Tumorbank
Deutsche Krebsforschungszetrum
Institut für Experimentelle Pathologie (dkfz)
Im Neuenheimer Feld 280
Postfach 101949
D-6900 Heidelberg 1
Feberal Republic of Germany
Telephone: (06221) 4841
Telex: 461562 DKFZ D

The tumour bank of the German Cancer Research Centre is another development; it produces a catalogue of all the tumour cell models available in Germany with details from nearly one hundred scientists in West Germany in addition to the tumour bank itself.

Hungary

National Collection of Agricultural and Industrial Micro-organisms (NCAIM)
Department of Microbiology
University of Horticulture
Somloi ut 14-16
H-1118 Budapest
Hungary
Telephone: (01) 665411
Electronic mail: TELECOM GOLD 751 DBI0185

The NCAIM is a collection with International Patent Depository Authority status which intends to carry cell lines.

Italy

Centro Substrati Cellulari
Instituto Zooprofilattico Sperimentale della Lombardia e dell'Emilia
Via A. Biandri 7
I-25100 Brescia
Italy
Telephone (030) 2290248
Telex: 305381 IZBS I
Fax: (030) 225613
Current holdings: 160 cell lines and hybridomas

The Centro Substrati Cellulari was established in 1979 as a tissue culture reference laboratory within the Instituto Zooprofilattico Sperimentale della Lombardia e dell'Emilia. The main activity of the laboratory is the propagation of cell cultures and study of the susceptibility by some cell lines to virus infection. Culture supply developed from these activities.

Japan

New developments are occurring in Japan and collections are being established in many centres.

Institute for Fermentation, Osaka
17-85, Juson-honmachi 2-chome
Yodagawa-ku
Osaka 532
Japan
Telephone: (06) 302-7281
Current holdings: 45 cell lines

The IFO began its collection of animal cell lines in 1984 and has a policy of collecting lines of special interest. Most cell lines are available without restriction. It is recognised as a Depository for Patent purposes by the European Patent Office. In addition, the collection offers safe deposit facilities.

Specialist collections

Japanese Cancer Research Resources Bank (JCRB)

The establishment of the research resources bank was part of a comprehensive 10-year strategy for cancer control begun in 1984, and comprises both a cell and gene bank. The organisation is funded by the Japan Shipbuilding Foundation and has a Cell Bank Management committee made up of members from the collaborating laboratories, the Ministry of Health and Welfare and the Foundation for the Promotion of Cancer Research. The main areas of interest are divided into JCRB-cell and JCRB-gene.

JCRB-Cell:

National Institute of Hygenic Sciences
Kami-Yoga
Setagaya-Ku
Tokyo
Japan
Telephone: (03) 700-1141
Fax: (03) 707-6950
Current holdings: 300 animal cell lines in eight constituent laboratories

The collaborating laboratories are:

(a) Food and Drug Safety Centre
Hatano Institute
Department of Cell Biology
Hatano
Kanagawa 257

(b) Toyko Metropolitan Institute of Gerontology
Itabashi-ku
Tokyo 173
Japan

(c) Institute of Medical Sciences
Tokyo University
Laboratory of Cancer Cell Research
Minato-ku
Tokyo 108
Japan

(d) Kihara Institute for Biological Research
Yokohama City University
Yokohama 232
Kanagawa 257
Japan

(e) Radiation Biology Centre
Kyoto University
Kyoto 606
Japan

(f) Cancer Institute
Okayama University
Division of Pathology
Okayamaken 700
Japan

(g) Radiation Effect Research Foundation
Department of Radiobiology
Hiroshima 730
Japan

At present, distribution of cell lines is primarily within Japan, but distribution world-wide is planned.

JCRB-Gene:
The role of this organisation is to distribute DNA derived from human cells.

National Institute of Health
2–10–35 Kami-Osaki
Shinagawa-ku
Tokyo 141
Japan
Telephone: (03) 444–2181
Fax: (03) 446–6286

The collaborating laboratories are:

(a) Department of Biochemistry
Faculty of Medicine
University of Tokyo
Tokyo 113
Japan

(b) Division of Cellular Biology
Institute for Molecular and Cellular Biology
Osaka University
Osaka 565
Japan

(c) Research Laboratory for Genetic Information
Kyushu University
Fukuoka 812
Japan

(d) Department of Biochemistry
Kanazawa Medical University
Ishikawa 920-02
Japan

(e) Department of Congenital Abnormalities Research
National Children's Medical Research Center
Tokyo 154
Japan

RIKEN Gene Bank
3-1-1 Yatabe
Koyadai, Tsukuba Science City 305
Japan
Telephone: (02975) 4-3 6 11
Fax: (02975) 4-2616
Current holdings: 50 cell lines

Another new development in cell banking is the opening of the RIKEN Gene Bank under the direction of the Science and Technology Agency. Both cell cultures and DNA are stored in a specialist laboratory at the Tsukuba Life Science Centre, near Tokyo. This bank also offers safe deposit facilities.

Patent Micro-organism Depository
Fermentation Research Institute
Agency of Industrial Science and Technology
1-1-3 Higashi
Yatabe
Tsukuba Science City 305
Japan

This bank is available for deposit of both animal and plant cell lines.

Ministry of Agriculture, Forestry and Fisheries Gene Bank
National Institute of Agriculture, Forestry and Fisheries
2-1-2 Kannondai
Yatabe
Tsukuba Science City 305
Japan

This is a gene bank for the deposit of a wide range of life forms, mainly seeds but also including animal cells.

United Kingdom

European Collection of Animal Cell Cultures (ECACC)
PHLS Centre for Applied Microbiology & Research
Porton Down
Salisbury SP4 0JG
UK
Telephone: (0980) 610391
Telex: 47683 PHCAMR G
Fax: (0980) 611315
Electronic mail: TELECOM GOLD 75: DBI0222
Current holdings: 1000 cell lines and hybridomas

The European Collection of Animal Cell Cultures was established in 1984 as a joint venture between the Public Health Laboratory Service and the UK Department of Trade and Industry.

This is the first animal cell service collection established within Europe. It has commenced the task of accessioning cell lines from many diverse laboratories in Europe, thus increasing availability world-wide of their unique resources. Work has also commenced on an enterprise funded by the Commission of the European Communities to establish a European Human Cell Bank. This consists of a well documented bank of B-lymphoblastoid cell lines prepared by Epstein–Barr Virus transformation from patients and their families with known genetic disorders.

The ECACC is an International Patent Depository Authority under the Budapest Treaty. It offers safe deposit facilities, cell line characterisation and quality control services and has capabilities for large-scale culture.

United States of America

American Type Culture Collection of Cell Lines
12301 Parklawn Drive
Rockville
Maryland 20852
USA
Telephone: (301) 881–2600
Telex: 908768 ATCC ROVE
Electronic mail: BT TYMNET 42:CDT0004
Current holdings: 2650 cell lines and hybridomas

The Collection of Cell Lines was established through the cooperative efforts of the ATCC, advisory committees and a group of collaborating laboratories. The cooperative programme was sponsored initially by the National Cancer Institute (NCI) in response to an urgent need in many areas of biological and medical research for authenticated and contaminant-free human and other animal cell cultures. Subsequently, the National Institutes of Health, the Division of Research Resources, the National Institute of Dental Research, the National Heart, Lung and Blood Institute, the NCI and the National Institute of Allergy and Infectious Diseases have offered contracts and grants which permitted expansion of the various subdivisions within the collection and the inclusion of hybridomas.

The collection now contains some 2650 characterised cell lines and hybridomas derived from approximately 75 different species, including 200 human skin fibroblast lines derived from apparently normal individuals and from patients with various disease states (including genetic disorders). All are preserved in liquid nitrogen. The reference cells are widely distributed. In addition to its role as a national repository of cell lines the collection also functions as the WHO International Reference Center for Cell Cultures.

The ATCC is an International Patent Depository Authority under the Budapest Treaty, offers safe deposit facilities and extensive cell line characterisation services.

National Institute of General Medical Sciences (NIGMS)
Human Genetic Mutant Cell Repository and National Institute on Aging Cell Culture Repository (NIA)
Corriell Institute for Medical Research
Copewood Street
Camden
New Jersey 08103
USA
Telephone: (609) 966-7377

Both NIGMS and NIA are divisions of the National Institutes of Health. The NIGMS Human Mutant Cell Repository was established in 1972 and now has over 3700 cultures which are available for human genetic studies. The collection includes apparently normal human fibroblasts, biochemically mutant human fibroblasts, chromosomally aberrant human fibroblasts, virus-transformed fibroblast cultures, apparently normal human lymphoblasts, biochemically mutant human lymphoblasts, chromosomally aberrant human lymphoblasts, human amniotic cell cultures and animal cell cultures including hybridomas and somatic cell hybrids.

The NIA Aging Cell Culture Repository was established in 1974 and has over 650 cultures which are available for studies into ageing. It includes well characterised human diploid cell cultures, human skin fibroblasts, differentiated cell cultures, cell cultures from patients with disease relevant to ageing and some non-human cell cultures.

The collection also offers services in mycoplasma detection and eradication techniques.

1.4 New developments in animal cell culture collections

The Bundesminister für Forschung und Technologie in the Federal Republic of Germany plans to fund a new animal cell culture collection as part of an expanded German Type Culture Collection to be established in Braunschweig. This collection will be closely linked to the existing European Collection of Animal Cell Cultures in the UK.

1.5 References

Barile, M. F., Hopps, H. E., Grabowski, M. W., Riggs, D. B. & Del Giudice, R. A. (1973). The identification and sources of mycoplasmas isolated from contaminated cultures. *Ann. NY Acad. Sci.* **225**, 252–64.

Gartler, S. M. (1968). Apparent HeLa cell contamination of human heteroploid cell lines. *Nature* **217**, 750–1.

Harris, N. L. Gang, D. L., Quay, S. C., Poppema, S., Zamecnik, P. C., Nelson-Rees, W. & O'Brien, S. J. (1981). Contamination of Hodgkin's disease cell cultures. *Nature* **289**, 228–30.

Hay, R. J. (1984a). Problems of specificity from the cell banking perspective: Colon cell lines at the American Type Culture Collection. In *Progress in Cancer Research and Therapy*, Vol. 29, ed. S. R. Wolman and A. J. Mastromarino, 3–21. New York: Raven Press.

Hay, R. J. (1984b). Banking and strain data of cell cultures. In *Uses and Standardization of Vertebrate Cell Cultures, In Vitro* Monograph 5, pp. 215–24. Tissue Culture Association.

Hay, R. J. (1985). *American Type Culture Collections Quality Control Methods for Cell Lines*. Rockville, Md.: ATCC.

Hay, R. J. (1986). Preservation and characterisation. In *Animal Cell Culture, a practical approach*, ed. R. I. Freshney, pp. 71–112. Oxford: IRL Press.

Hay, R. J., Williams, C. D., Macy, M. L. & Lavappa, K. S., (1982). Cultured cell lines for research on pulmonary physiology available through the American Type Culture Collection. *Am. Rev. Respir. Dis.* **125**, 222–32.

Hopps, H. E., Meyer, B. C., Barile, M. F. & Del Giudice, R. A. (1973). Problems concerning 'noncultivable' mycoplasma contaminants in tissue culture. *Ann. NY Acad. Sci.* **225**, 265–76.

Lavappa, K. S. (1978). Survey of ATCC stocks of human cell lines for HeLa contamination. *In Vitro* **14**, 469–75.

McGarrity, G. J. (1982). Detection of mycoplasmal infection of cell cultures. *Adv. Cell Culture* **2**, 99–131.

Nelson-Rees, W. & Flandermeyer, R. R. (1977). Inter- and intraspecies contamination of human breast tumor cell lines HBC and BrCa5 and other cell cultures. *Science* **195**, 1343–44.

Nelson-Rees, W., Daniels, D. W. & Flandermeyer, R. R. (1981a). Human embryonic lung cells (HEL–R66) are of monkey origin. *Arch. Virol.* **67**, 101–4.

Nelson-Rees, W., Daniels, D. W. & Flandermeyer, R. R. (1981b). Cross-contamination of cells in culture. *Science* **212**, 446–52.

Nelson-Rees, W., Flandermeyer, R. A. & Hawthorne, P. K. (1974). Banded marker chromosomes as indicators of intraspecies cellular contamination. *Science* **184**, 1093–6.

Scherer, W. F. (1962). Program on characterization and preservation of animal cell strains. *Natl. Cancer Inst. Monogr.* **7**, 3–5.

Stevenson, R. E. (1963). Collection, preservation, characterization and distribution of cell cultures. *Proc. Symposium on the Characterization and Uses of Human Diploid Cell Strains*, p. 417. International Association of Microbiological Societies.

Stulberg, C. S., Coriell, L. L., Kniazeff, A. J. & Shannon, J. E. (1970). The animal cell culture collection. *In Vitro* **5**, 1–16.

2

Information resources

L. BLAINE, M. I. KRICHEVSKY and H. SUGAWARA

2.1 Introduction

Microbiologists and cell biologists are faced with consideration of exponential growth in their laboratories on a daily basis. As users of a chapter on information resources for biotechnology they are exposed to a double dose of exponential growth. First, the explosion of information technology itself is due to the massive amounts of computing power available at ever diminishing cost. In turn, a population of computer aware and computer literate microbiologists represent a growing demand for more sophisticated access to modern information technology. The community of information technologists in concert with microbiologists are responding to this demand with a multiplicity of initiatives using various strategies.

The resulting activity induces feelings of inadequacy in the authors of such chapters as this, since at the moment of delivery to the editors the information is out of date. Resources previously known only by rumour are tested. Simple facilities being tested as pilot projects are quickly made available to the community. Local data banks open their doors to regional and even world-wide participation. Databases on databases spring up because of the need to discover available resources. The net result is an ever increasing base of information resources for biotechnologists.

In some cases, useful resources fall by the wayside, as have at least two of the resources listed. They have been discontinued in the interval between the first and present versions of this chapter. The root cause of such discontinuing of effort is lack of appreciation by the initial funding bodies for the complexity and time scale involved in database initiatives of this sort.

While the information about information presented in this chapter is out of date as soon as it is written, the resources described are most likely to be improved and be more useful than the descriptions indicate. For information on new developments, the listed resources should be contacted.

2.2 Information needs

The need of the biotechnologist for widely disparate categories of information is a consequence of the varied nature of the tasks required to design, develop, and consummate a process. The biotechnologist must find or develop genotypes of the required composition, discover the conditions for expression of the desired phenotypic properties, maintain the clones in a stable form, and describe all of these parameters in a fashion understandable to peers. Most of the categories of information required will be outside the expertise of any one individual. Thus, a panel of experts representing all disciplines involved must be assembled or access to diverse databases must be achieved. The library or publicly accessible databases can lead to the required information sources which range from the traditional scholarly publications to assemblages of factual or primary laboratory observations. This chapter presents an overview of the kinds of information available and the mechanisms to access them. In particular, it will concentrate on information resources for finding material with the desired attributes, be they taxonomic, historical, genetic, or phenotypic.

2.2.1 *Interdisciplinary information sources*

The personal training and professional experience in the particular narrow field in which they work allows microbiologists and cell technologists to perform many daily tasks without reference to outside sources of information. However, the interdisciplinary nature of the practice of biotechnology forces the use of navigational aids to the knowledge base of unfamiliar areas. This point is demonstrated by the observation that approximately 80% of the enquiries to the CODATA/IUIS Hybridoma Data Bank (see below) are from persons who are not immunologists. It follows that a successful database resource should be designed with the interdisciplinary users in mind as they often will form the largest segment of the user population.

The main pathways to locating an existing source of strains or cell lines with the properties that will be useful in the projected process are through records of the primary observations of properties or through derived information such as taxa or strain designations. In either case, the desired result is one or more strain designations and instructions on where to get cultures. Even though the desired end result is the same with both pathways, the mechanisms for recording and disseminating the information are usually, but not necessarily, quite different.

Culture collections which have a stated mission of providing service to a public user community, especially through distribution of cultures outside their host institution, tend to use a taxonomic orientation in that their records are commonly kept as discrete strain descriptions, often one strain to a page. The whole strain description is easily read while comparisons among strains are difficult.

Culture collections serving predominantly as local institutional repositories of strains for research, teaching, or voucher specimens for archival storage, tend to rely on primary observation data kept in tabular form with the attribute designations as column headings and the strain designations as row labels. In contrast to the previous case, comparisons among strains are easily made, while assembling a complete strain description may require following the row designation for the strain through multiple tables.

Even now, most service collections use traditional paper-based data management methods rather than computers. Within a few years, the majority of collections will be using computers as their main data-handling tool. The functional distinctions between these alternative forms of data organisation blur with the use of computers, but can still be a factor if good information management practices are not followed.

2.2.2 *Where to get microbial strains or cell lines*

The ultimate source for clones with desired properties is isolation from nature. Indeed, large efforts have been mounted to find strains with desired properties such as production of antibiotics. These efforts are feasible when a good screening procedure, such as zones of clearing around a colony, is available. Even then, the effort is labour-intensive with resulting high costs.

The existence of culture collections with data on the characteristics of the holdings makes possible enormous savings when appropriate cultures are available. However, the data must be available to the process developer with reasonable ease. The best source for cultures will often be from the collection with the most available data rather than the most complete selection of strains or cell lines with likely sets of characteristics.

2.2.3 *How to get microbial strains or cell lines*

The pathway used in obtaining access to the required data for finding desirable strains starts with the same foci as the collections themselves. A taxonomic strategy or primary observation strategy may be used. A taxonomic search strategy for strains producing higher concentrations of a particular material (e.g. penicillin, riboflavin, ethanol) might well begin with asking service culture collections for all strains of *Penicillium notatum, Ashbya gossypii, Saccharomyces cerevisiae* in their collections and screening them for level of production. This method of searching requires the searcher to know which taxa are likely to have the desired attributes.

A primary observation search strategy for cultures with the ability to degrade a particular material might well start with asking for all cultures that degraded that material and had the growth characteristics that were desirable under the projected process conditions. The question to the collection might be 'Could you provide me with the characteristics of all your cultures able to use hexadecane while growing aerobically at 25 °C?'. This method of searching requires no taxonomic knowledge on the part of the searcher. Clearly, the format of data storage in the collection would markedly affect the relative ease of answering these questions.

2.2.4 *Strain or cell line data*

The traditional classes of data used to describe strains in culture collections still predominate. These include morphological, physiological, biochemical, genetic, and historical data. Clearly, these classes will form the basis for all future collection data as well. The current emphasis on biotechnology places new demands on the informational spectrum desired of microbial and cell line collections. In addition to the requests for taxa with specific properties, information is also requested on potential utility of strains in biotechnological processes and the actual or potential hazards arising from their use. These ancillary but important data are sparse in their availability, but some service collections with a concern for the needs of industry are collecting data of attributes specifically useful in biotechnological processes such as temperature tolerances, behaviour in fermenters, stability of monoclonal antibodies under adverse conditions, toxicity of products, and pathogenicity for unusual hosts.

Physiology. The bulk of information included in publicly accessible databases will be physiological and biochemical. Most queries will be on what the strains or cell lines may be capable of doing with respect to the desired process, and the databases will be heavily weighted towards containing this kind of information.

While printed compendia of physiological and biochemical data are technically possible, they are rare and, in view of the expense of publishing such volumes, will continue to be rare. Rather, these kinds of data are more reasonably compiled in bulk and disseminated through computers.

Morphology. The detailed morphological data held by most culture collections are generally of less interest than details of physiological data for process development. In various broad classes of organisms (fungi, protozoa, and algae) morphology may be critically important for taxonomy and identification. These same detailed attributes usually have little bearing on the conduct of most processes. However, some basic morphological information will be of fundamental importance in process engineering. For example, cell size knowledge is needed if filtration is a part of the process. Likewise, use of filamentous strains will require more energy for stirring fermenters than non-filamentous forms. Additionally, knowledge of sexual reproduction or fusion attributes is critical if hybridisation or strain improvement programmes are to be carried out.

Publicly accessible electronic databases used in searches will, with few exceptions, have only basic or general morphological information because of the costs associated with printing or storage of such information.

Industrial. Data on the use of particular strains or cell lines for specific processes is a frequently requested category of data. Such data exist, but in a widely scattered, uncoordinated fashion. They are contained in the open literature, collection catalogues, patent disclosures, and other more obscure repositories. The same is true of related data on the behaviour in the processes themselves. There are few biological data compendia equivalent to the materials properties databases available to the engineers in chemistry, metallurgy, and similar disciplines. This situation is partly due to the nature of biological material and partly due to the development of high technology in biology later than in the other disciplines, in spite of the ancient history of biotechnology.

Hazard. Risk assessment of biotechnological activities concerns governmental regulatory bodies throughout the world. All of the concerns are environmental and may be in such forms as a specific disease of humans, animals, or plants, an undesirable imbalance in the environment, or production of an undesirable product.

The data available to answer queries in the area of risk assessment are quite sparse. The most common category of useful data is pathogenicity. This is largely a strain-specific phenomenon, the degree of virulence varying from strain to strain, especially on serial propagation. Less commonly available are data describing strain persistence in various environments and toxicity of products and, rarest of all, perhaps data to predict the effect of the introduction of strains into new environments.

2.2.5 *Taxonomic data*

The traditional and important method of constructing databases in service culture collections is with taxonomic orientation. The storage of the data and their public presentation considers the strains first as representatives of their taxa. Further, the data given in the description of each strain in the catalogue of collection holdings are sparse as they depend on the assumption that the reader knows, or is adept at finding, the usual attributes of the taxon. Many clinical microbiology laboratories only save the phenotypic data on antibiotic resistance patterns and the putative name of the isolates; the data used to decide the name are discarded. The records may indicate that the isolate is 'atypical' without any indication of the attributes that led to this description.

Given the traditional organisation of culture collection catalogues, the most common questions asked of service culture collections are likely to be on specific attributes of strains listed in the catalogues, or the companion question on the availability of strains with specific attributes. Indeed, experience shows that service collection personnel quickly develop great skill in searching the collection database by any and all means available to answer such questions from the community. In turn, the community learns that the collections are a valuable source of various kinds of information beyond that contained in the catalogues.

The taxonomic orientation is natural in managing diverse culture collections and determining the boundaries of interest for many speciality collections. Within a large collection, curator responsibilities are usually assigned along taxonomic lines.

Historically, many service collections were established for studying or supporting the study of taxonomy. This essential function still underlies a great deal of service collection activity.

Many of the laws and regulations concerning shipment, hazard, standards, laboratory safety, patents, and use of biological material are stated all or in large part in taxonomic terms. The US Federal Register refers to strain and species in defining which strains are to be used as standards for antibiotic testing. The attributes of these strains are not listed. Similar regulatory documents exist for other parts of the world.

2.2.6 *Regulations*

The development of regulations for efficient and safe use of biotechnological processes to the common good is heavily dependent on adequate data of diverse types. Unfortunately, databases do not exist to allow detailed appraisal of the potential hazard on a case-by-case basis. The use of taxonomic levels to evaluate and regulate is contraindicated by the very nature of the process of establishing taxa. The only solution to this dilemma requires the gathering and dissemination of appropriate data.

2.3 Information resources

An informal infrastructure of information gatherers, managers, and disseminators exists to answer questions on the practice and regulation of biotechnology as it relates to microbial strains and cell lines. It forms a very useful resource in spite of its informality. Further, a number of initiatives are under way which aim to manage this support system for biotechnology in a more formal and complete fashion.

The infrastructure exists independently of the practice of biotechnology since the same needs for information transfer are basic to all of microbiology and cell biology. Collection holdings are raw material for these sciences and new information initiatives are stemming from the application of molecular biology and advances in biotechnology. These developments are merely refining the information pathways that evolved before modern genetic engineering focused the public eye on one of the oldest forms of manufacturing, the use of biological materials in processes.

The historical sequence of development of the infrastructure started with the culture collections, proceeded with catalogue production, followed by collections of strain or cell line data in computers and, most recently, the creation of national, regional, and finally international data services.

The ultimate source of data needed by the biotechnologist is the laboratory records of the collections. All other elements of infrastructure function to make these data accessible. The entry points to the pathways at all levels are the same as those considered previously: taxonomic or by detailed pattern of attributes. A combination is possible. The question, 'Do you have a pseudomonad that degrades hexadecane?' will eliminate all yeasts that have the same ability.

2.3.1 *Culture collection catalogues*

Service collections publish catalogues describing their available strains or cell lines to inform the public of their holdings and the salient properties. A secondary effect is to reduce some of the labour overhead involved in answering questions from the public. A number of nations (such as Japan, People's Republic of China, and Brazil) have prepared combined catalogues, eliminating the need to obtain and consult multiple catalogues.

By the very nature of printed catalogues, they function imperfectly since the information describing each strain or line is limited. Only one or at best a few attributes can be indexed so that detailed searching is impossible. Because of the positive accessioning policies of the service collections, catalogues are out of date upon publication. In general, the taxonomic entry route is served well, but the attribute route is necessarily left to follow-up questions to the collection itself. The most important deficiency is that only a very small proportion of the world's collections publish catalogues at all, since provision of a public service is not their prime function.

Where catalogues do exist, they form an important resource for the biotechnologist. The information they contain is carefully presented. If a taxon is known, much time can be saved by consulting a catalogue for availability. Often, valuable ancillary information on use, literature references, propagation conditions, or patents is included in the catalogue. Finally, the staff of the collection listed in the catalogue can be contacted for further information on their holdings or as entry into the rest of the informal information services of the collection community.

2.3.2 *Individual collections*

Culture collections of importance in biotechnology are not limited to the recognised service collections. In many cases, the biotechnologist must have access to a detailed collection of strains or cell lines with the final selection for use in the process decided by personal comparative testing. Service collections may serve in this respect, but because of their usually broad nature cannot always maintain the detailed holdings of the personal research or survey collection.

The mechanics of obtaining information from primary records of individual collections may not be simple. The curator must scan tables or individual strain or cell line records to match the pattern of the query. The process is faster for those collections which keep their records in a computer. While computer-aided searching is faster, considerable time is required of collection personnel to search on a query by query basis. Time spent in searching the database to answer public queries may be considered a normal and reasonable function by the service collections or a burdensome chore by collections with other basic missions. Either way, such searches can represent a considerable overhead on staff time.

2.3.3 *Strain or cell line data compendia*

The availability of computers for both record keeping and data analysis has resulted in compendia of strain or cell line data in locations other than the collections. Hospitals have contributed antibiotic resistance data to large-scale surveys of changes of resistance plasmid distributions in bacterial populations. Numerical taxonomists assemble large databases of strain data in the course of their activities, often containing data contributed by other research workers. Ecologists and public regulatory agencies conducting surveys of the environment, including habitats such as soils, waters, foodstuffs, and wild and cultivated plants, often amass considerable amounts of data which are installed in some computer resource for management and analysis. The result is that the data describing the strains and cell lines reside in a different location from the cultures.

Arising from all this activity is a body of curators of data in support of the curators of the biological material. The relevant information specialists may be associated with a collection, and where such data management exists, the biotechnologist's search is immensely facilitated. Since the taxonomic designation is managed in the computer in the same way as any attribute of the strain, the entry into the database can be taxonomic or by attribute pattern with equal facility.

The problem of finding the strains of interest is largely reduced to the problem of finding the databases themselves.

2.3.4 *National, regional, and international data resources*

Scientists and technologists band together in organisations focused on their disciplines in order to exchange information. Many of these scientific and technical societies also become providers of services to their members and the public community. Some resemble guilds or unions in providing advocacy for improving conditions for their members. Recently, societies have come full circle in that they are providing, directly or through advocacy, informational resources in the form of publicly available databases.

Since most societies are national, the earliest of these database efforts were national as well. In some disciplines national efforts were deemed so valuable that they became international in use. The Chemical Abstracts Service of the American Chemical Society is a well known example. Geopolitical regions have their counterparts in scientific and technical activities. The most notable of these regional efforts is within the European Economic Community with a ripple effect to other countries in Europe.

Either by combining regional resources or by direct international efforts, world-wide database resources are being established in many disciplines. Some stand alone and others are distributed in networks.

Informational resources describing culture collections and their holdings are distributed through all three levels as are the organisations concerned with sponsoring such resources.

National and regional organisations concerning culture collections. Microbiologists and cell biologists having interests in culture collections have banded together in national federations for culture collections. The countries that have such national federations are listed and further described in Chapter 8. The countries are Australia, Brazil, Canada, China, Czechoslovakia, Japan, Korea, New Zealand, Turkey, United Kingdom, and United States of America. Japan, Brazil and the UK are most actively pursuing national databases on collection holdings. Others are being discussed or planned. The Japanese database is designed to contain information from all types of microbial collections. The Brazilian and UK systems are initially designed to concentrate on service collections. The European Community has recently established a regional information service; the east and central European countries are actively planning networks on national and regional lines. Other databases are being discussed or planned and more details of these systems are given below.

Brazil

The third edition of the *Catalogo Nacional de Linhagens* was released by the Fundação Tropical de Pesquisas e Tecnologia 'André Tosello' in Campinas, Brazil in 1989. The species names and designations of strains and cell lines held by Brazilian collections are listed in the catalogue. The catalogue includes bacteria, filamentous fungi, yeasts, protozoa, algae, animal cell lines, viruses, and miscellaneous microorganisms.

The on-line version of the Catalogue is available on the Base de Dados Tropical, which is installed on the Brazilian national information system, Embratel and is accessible through international telecommunication systems with the proper access codes and billing arrangements. In addition to catalogue information, the BDT contains a directory of specialists and research projects in applied microbiology in Brazil.

Contact:
Fundação Tropical de Pesquisas e Tecnologia 'André Tosello'
Rua Latino Coelho, 1301
13.000 Campinas, SP
Brazil
Telephone: (0192) 42-7022
Electronic mail: BT TYMNET 42:CDT0094

European Culture Collections' Organisation (ECCO)

ECCO is an active group comprised of representatives of major service collections in countries that have microbiological societies affiliated with the Federation of the European Microbiological Societies (FEMS). All the ECCO collections are also affiliated with the World Federation for Culture Collections. Forty-one collections are represented from Belgium, Bulgaria, Czechoslovakia, Finland, France, Federal Republic of Germany, German Democratic Republic, Greece, Hungary, Italy, Netherlands, Norway, Poland, Portugal, Spain, Sweden, Switzerland, Turkey, United Kingdom, Union of Soviet Socialist Republics and Yugoslavia.

ECCO was established in 1982 to collaborate and trade ideas on all aspects of culture collection work. Since service collections inherently are organised repositories of information as well as cultures, ECCO is a valuable resource for finding information of interest in biotechnology. Each of the collections produces a catalogue of holdings. Such catalogues often hold information beyond the listing and description of the cultures held. Also, the curators are likely contacts for other, non-service, collections in their countries.

As intercollection communication pathways grow and become formalised within and between ECCO countries, the prospects for an electronic communication network encompassing all these countries grows as well. Such a network is being actively discussed among its members at this time. ECCO members are collaborating with the Information Centre for European Culture Collections (see below).

Contact:
USSR Collection of Microorganisms
Institute of Physiology and Biochemistry of Microorganisms
Academy of Sciences
Pushchino-na-oke
142292 Moscow Region
USSR
Telephone: (095) 2316576
Fax: (095) 9233602

Japan Federation for Culture Collections (JFCC)
JFCC has promoted cooperation among culture collections and individuals in Japan as well as internationally since 1951. In 1953, the JFCC started collecting data on the holdings in Japan and completed the catalogue which listed about 22 000 strains from 144 research institutions. The strains in the list were reidentified by a project team organised by Professor Kin'ichiro Sakaguchi of the University of Tokyo, resulting in the publication of a series of JFCC catalogues in 1962, 1966, 1968, and the most recent in 1987. The JFCC produces a Bulletin that is published regularly.

Contact:
JFCC
NODAI Research Institute Culture Collection
Tokyo University of Agriculture
1–1–1 Sakuragaoka, Setagaya-ku
Tokyo 156
Japan

Microbial Resource Centres (MIRCENs)
Information on Microbial Research Centres (MIRCENs) will be found in Chapter 8. These centres are found in both developed and developing countries. Each has its special focus of interest, for which it acts as a regional centre.

National and regional data resources

Information Centre for European Culture Collections (ICECC)
In 1986 the Commission of the European Communities provided funds under its Biotechnology Action Programme for the establishment of an information centre for European culture collections. An independent organisation, the centre is a useful source of information on all aspects of culture collection activity in Europe and beyond.

Its mission is to hold information on holdings and specialised services, such as identification or patent services, available from the service collections in Europe. In addition it promotes the services of the collections at conferences and exhibitions, plans training courses and is able to act as the Secretariat for European culture collections.

The MiCIS database (see below) was transferred to the ICECC in 1989 and is now available on-line. It is necessary to register with the ICECC to obtain access to the database. It is planned that this database will form the nucleus for a centralised European database holding data on individual strain properties. A gateway between MSDN (see p. 41) and ICECC operates, giving users the opportunity to access the centre either direct or through the MSDN network.

Contact:
Information Centre for European Culture Collections
Mascheroder Weg 1b
D–3300 Braunschweig
Federal Republic of Germany
Telephone: (0531) 618715
Fax: (0531) 618718
Electronic mail: TELECOM GOLD 75:DBI0274

Institute for Physical and Chemical Research
In Japan, RIKEN carries out international activities, such as being the host institution for the World Data Center on Collections of Microorganisms and a node of the Hybridoma Data Bank. These activities are carried out by the Life Science Research Information Section (LSRIS) and Japan Collection of Microorganisms (JCM).

On a national level, LSRIS has developed the National Information System of Laboratory Organisms (NISLO), a directory of Japanese collections and their holdings. The NISLO covers microorganisms, animals, and plants. In the case of microorganisms, the LSRIS closely cooperates with the JCM.

The following information and services are currently available from LSRIS:

(1) the number of laboratory animals used in Japan and their scientific names;
(2) microorganisms maintained in the member collections of JFCC;
(3) algae maintained in culture collections in the world;
(4) identification of deciduous trees;
(5) fundamental references for cell lines widely used in Japan;
(6) bibliographical information for plant tissue and cell cultures.

Contact:
LSRIS
RIKEN
2-1Hirosawa, Wako
Saitama 351-01
Japan
Electronic mail: BT TYMNET 42:CDT0007

Microbial Culture Information Service (MiCIS)

One of the first (along with Japan) national data efforts is that of MiCIS in the UK. This initiative is oriented towards providing public access to primary observation data while retaining the taxonomic orientation for computer entry of the data. The initiative is a cooperative effort between the UK Department of Trade and Industry (DTI) and the UK Federation for Culture Collections. The following description is quoted from a joint statement of purpose between MiCIS and MINE (see below).

> This database has been developed by the Laboratory of the Government Chemist (LGC) on behalf of DTI following consultation with industry. MiCIS initially contains all data currently available on strains including catalogue information, hazards, morphology, enzymes, culture conditions, maintenance requirements, industrial properties, metabolites, sensitivities and tolerances. Further categories of

> information will be added as the system develops. MiCIS will contain data from all UK national culture collections and discussions are currently taking place to include data from private and other European collections.

The funding for the MiCIS effort was terminated in 1989. The database is not being expanded or actively maintained by the culture collections at this time. However, the database has been transferred to and is available on-line from the Information Centre for European Culture Collections (see above). It is also available on-line through the Microbial Strain Data Network (see Section 3.3 below).

Contact:
Information Centre for European Culture Collections (ICECC)
Mascheroder Weg 1
D–3300 Braunschweig
Federal Republic of Germany
Telephone: (0531) 618715
Fax: (0531) 618718
Electronic mail: TELECOM GOLD 75: DBI0274

Microbial Information Network Europe (MINE)
Within the European Economic Community (EEC) the regional activities are primarily focused within the programmes of the Commission of the European Community (CEC). They fund a number of initiatives, either as sole efforts or as collaborations when the initiative has a scope beyond the confines of the EEC. One such initiative within the EEC is MINE. MINE has the traditional taxonomic orientation of the service collections. The UK node has provided the following description of MINE in a joint (with MiCIS) statement of purpose

> This EEC database is a computerised integrated catalogue of culture collection holdings in Europe and is prepared as a part of the EEC Biotechnology Action Programme. CMI [Commonwealth Mycological Institute], in collaboration with CABI [Commonwealth Agricultural Bureau International] Systems Group, is to act as the UK node, in parallel with national nodes being developed in The Netherlands, Germany, and Belgium; discussions with Portugal and France are currently under way.
>
> . . . it will not include full strain data but only a minimum data set.

> Once enquirers locate a culture, they will then be referred to the collection concerned, or to national strain data centres . . . for any detailed strain information if this is needed.
>
> Initially, enquirers will contact MINE nodes by mail, telex, or telephone. At a later stage, on-line services may also be available on subscription.

Some of the above statement may need to be modified with the passage of time. A hard copy integrated catalogue and minimum dataset are deferred. Rather, MINE has become a cooperating group of collections. Various collections within MINE have established their own databases using similar software and agreed upon fields. Others have databases that are not directly compatible with others. Further, some of the collections have advanced to the point of having independent on-line systems available for public enquiry. There is agreement within MINE for support of a centralised database with full data set of strain characteristics, possibly within the framework of the Information Centre for European Culture Collections and based upon the MiCIS database now converted to the MINE format and maintained at the Centre.

Contact:
Information Centre for European Culture Collections (ICECC)
Mascheroder Weg 1
D-3300 Braunschweig
Federal Republic of Germany
Telephone: (0531) 618715
Fax: (0531) 618718
Electronic mail: TELECOM GOLD 75:DBI0274

Nordic Register of Microbiological Culture Collections

In 1984, the Nordic Council of Ministers initiated support for a Nordic Register of Culture Collections encompassing Denmark, Finland, Iceland, Norway, and Sweden. The Register's scope is inclusive of all sizes and functions of collections from small personal collections to large service collections. The first three years of development focused on strains of importance to agriculture, forestry, and horticulture. The building of the microcomputer-based database was an undertaking of the Department of Microbiology of the University of Helsinki, Finland. Development of software has been carried out in cooperation with the Nordic Gene Bank for Agricultural and Horticultural Plants in Alnarp, Sweden.

In 1987, the funding from the Council of Ministers ceased. For this reason, the Nordic Register is not functional.

International Resources

The primary focus for international culture collection information resources is through the International Council of Scientific Unions (ICSU) with headquarters in Paris. Various components of ICSU have current or potential initiatives relating to providing information of interest to biotechnologists. These include: Committee on Data for Science and Technology (CODATA), World Federation for Culture Collections (WFCC), International Union of Immunological Societies (IUIS), and International Union of Microbiological Societies (IUMS).

Committee on Data for Science and Technology (CODATA)

CODATA was established in 1966 by the International Council of Scientific Unions (ICSU) to promote and encourage the production and international distribution of scientific and technological data. Its initial emphasis was in physics and chemistry, but its scope has been broadened to data from the geo- and biosciences. CODATA is 'especially concerned with data of interdisciplinary significance and with projects that promote international cooperation in the compilation and dissemination of scientific data'.

The main activities of CODATA are carried out by Task Groups established for specific projects. Of special interest to biotechnologists are the biologically oriented Task Groups on the Hybridoma Data Bank, Microbial Strain Data Network, and Biological Macromolecules. The roles of these Task Groups vary. The first gives scientific guidance (policy), the second is advisory only, while the last is a coordinating body among existing sequence data banks.

The latest initiative of CODATA in biological information is the creation of a Commission on Terminology and Nomenclature in Biology. The purpose of the Commission is to locate existing standards of terminology or nomenclature in the disciplines of biology, publicise the existence of the standards, and encourage the development of standards where they are deemed lacking. The Commission has sought and obtained the cooperation of most of the constituent biologically oriented Unions of the International Council of Scientific Unions. While the development of standards is the responsibility of the Unions, the Commission will make the considerable experience of its members in developing computer-based vocabularies and databases in general available to the standard-setting bodies. The first such effort is in the area of viruses in collaboration with the International Committee on the Taxonomy of Viruses.

Contact:
CODATA Secretariat
51 Boulevard de Montmorency
75061 Paris
France
Telephone: (1) 45-25-04-96
Telex: 630553
Cable: ICSU PARIS
Fax: (1) 42-88-94-31
Electronic mail: TELECOM GOLD 75:DBI0010

Hybridoma Data Bank (HDB)

The multidisciplinary committee on Data for Science and Technology (CODATA) of the International Council of Scientific Unions (ICSU) and the International Union of Immunological Societies (IUIS), recognising the significance of hybridoma technology for all fields of biology, are sponsoring an international effort to collect and disseminate information on the products of this technology. This project, The Hybridoma Data Bank (HDB), is more appropriately termed, 'a Data Bank on Immunoclones and their Products'. The term 'immunoclones' was suggested as a subtitle for the HDB by Professor Alain Bussard, Secretary General of CODATA and one of the initiators of the Bank, because its scope includes all cloned cell lines and their immunoreactive products.

Information specialists at the American Type Culture Collection (ATCC), Japan's Institute for Physical and Chemical Research (RIKEN), the Centre European de Recherches Documentaires sur les Immunoclones in France (CERDIC) and the European Collection of Animal Cell Cultures (ECACC) in the UK are collecting and disseminating data on hybridomas and other cloned cell lines of immunological interest. These reagents have wide applicability and are used in scientific disciplines as diverse as clinical medicine, environmental biology, agriculture, and biological systematics. Because of this diversity, the HDB is of interest to much of the scientific community.

Laboratories around the world are generating immunoclones at a rapid rate. Many are finding it difficult to keep track of the data associated with these cell lines, even within their own laboratories. Therefore, the requirement for a centralised registry is widely recognised. The editorial policies of most scientific journals do not permit publication of reports on new hybridomas or other cell lines unless they include data on a significant new application for the products of these clones. This necessitates an alternative mechanism for information exchange. The HDB, a computerised central repository and information centre for hybridoma technology provides the scientific community with data which would be otherwise difficult to obtain. General categories of information stored in the HDB are as follows:

Developers of specific immunoreactive cell lines and products
Immunocyte donors
Immunising agents and procedures
Methods of conferring immortality
Immortal cell partners
Soluble immunoreactive products
Reactivities of monoclonal antibodies
Known non-reactivities of monoclonal antibodies
Uses and applications of immunoreactive products.
Availability and distribution of products and services

Specific types of queries which can be answered by a Data Bank search are discussed later.

The computerised data storage and retrieval system used to manage the hybridoma information has evolved into a versatile and unique 'hybrid' system which takes advantage of the attributes of both text based and binary or numeric systems. It continues to grow as new requirements are identified, and is serving as a model for databases now being developed by hybridoma technology. Therefore the project is truly international in scope.

Administrative structure. All technical and administrative activities of the HDB are overseen by a Task Group operating under the sponsorship of CODATA. The members of the Task Group, appointed by the Task Group Chairman, rotate periodically. All are internationally known as scientists with expert knowledge of hybridoma technology and its applications and/or data storage and retrieval systems.

The HDB Task Group has devised an overall operating structure for the Bank. This structure consists of Nodes and a Validation Center. The Validation Center is currently located at ATCC with all Nodes helping to support the cost of quality control and maintenance of a hierarchical list of preferred terms and synonyms.

Regular communication among the Nodes is handled principally through the CODATA MSDN Network on the BT TYMNET system. This network provides the means for electronic mail communication and file transfer among the Nodes.

There has been interest expressed in the establishment of Nodes in other areas of the world, such as Australia and China. India and Canada have recently joined the network and will become active partners in the collection and dissemination of data. It is hoped that more new Nodes can be established in the future when coordination and data exchange problems have been worked out.

Data collection and dissemination. Data are collected for the HDB at all three Nodes by direct solicitation from investigators generating immunoclones and from commercial distributors of immunoreactive products of cloned cell lines, such as monoclonal antibodies.

The number of searcheable records in the Bank is now approaching 15 000. It is estimated, however, that there are well over 50 000 hybridoma products now available for unrestricted use or collaborative research, so there is still much work to be done.

Methods of data collection vary slightly among the Nodes. Data Bank staff at the RIKEN have established an excellent communication network with Japanese immunologists. Data is submitted directly to the RIKEN on data reporting forms which have been circulated throughout the immunology community in Japan.

The US and European Nodes also collect data directly from the research community, but perhaps because of the more diffuse nature of this community in North and South America and in Europe, the network of contributors is not as strong. Therefore, it is necessary to supplement direct submissions with data from the literature. At the US Node, the Data Bank is growing in direct response to queries which come into the Bank. If a request comes in for information which is not found in the HDB, the staff perform literature searches and attempt to determine if and where such information exists. Data are coded directly from the literature. Authors are contacted to determine the availability conditions for their cell lines and products described in the literature and for unpublished information on other hybridomas.

Another rich source of data entered by the US and European Nodes is the 'commercial literature'. Catalogues and data sheets produced by vendors of monoclonal antibodies and other products of hybridoma technology are perused for new data. Commercial vendors are anxious to provide information to the HDB because it is an excellent marketing vehicle for their products.

Through interactions with major centres of hybridoma technology, it has been realised that many laboratories are struggling to computerise the in-house data being generated by this technology. The US Node has initiated a series of workshops on hybridoma data management. It is hoped that an increased awareness of the HDB and increased data sharing by investigators will result. As individual laboratories begin to computerise their data, using HDB format as a model, they can be transferred to the central Bank electronically. This eliminates the laborious process of reformatting and coding data from hand-written report forms.

Queries from the scientific community are accepted by each of the Nodes. Data Bank searches are performed by information professionals at the Nodes, and results, usually in the form of mailed printouts, are sent out.

The US has developed a prototype Directory to the HDB which is available on the Microbial Strain Data Network (MSDN). The on-line prototype is accessible from most countries of the world. It is a menu-driven system which can be easily used to obtain a listing of immunoclones in the HDB along with some basic information on characteristics and availability. Users of the prototype on-line database can communicate with all of the Nodes of the HDB via electronic mail to request further information. The system also allows for direct communication with vendors of monoclonal antibodies and on-line product ordering.

Data Bank search and retrieval. The HDB is designed for highly specific data retrieval. All of the data fields listed in Table 2.1 are searchable. Data are entered into these fields according to a hierarchical controlled vocabulary. Data are retrieved by entering the same controlled vocabulary terms or their corresponding numerals. Terms may be combined using standard Boolean operators (and, or, not) for specificity of retrieval.

Search strategy used for data retrieval depends on whether the numeric version of the HDB or the text version is being searched.

Table 2.1. *Fields and sub-fields used in the Hybridoma Data Bank*

Field Header Abbreviations

AU = Author (developer of cell line or product)
AD = Address of cell line or product developer
SO = Reference source
DI = Distributor of cell line or product
DE = Designation of individual cell line or product
IM = Immunogen
RM = Immunization route/immunization method
DO = Immunocyte donor
IP = Immortal partner
PR = Immortalization prodecure
PD = Type and classification of product
AS = Assay procedure used to determine immunoreactivity of product
RE = Reactant(s)
XR = Cross-reactant(s)
NR = Non-reactant(s)
RT = Reactants tested
AP = Use or applications of product
AV = Availability information
AB = Comments

Codes used within fields

G> = Genus species name of organism
CN> = Common name of organism
S> = Strain of organism
O> = Organ source
T> = Tissue source
CE> = Cell source
U> = Subcellular source
CD> = Cell line designation
HT> = Histocompatibility type
BT> = Blood type
PA> = Pathological state
D> = Developmental stage
AG> = Age
SN> = Substance name
MW> = Molecular weight
CC> = Classification code
B> = Biochemical composition
FS> = Fine specificity
PM> = Purification method
C> = Cell line information
P> = Soluble product information
YR> = Year line was established
VS> = Viral structure
%> = Percent reactivity
CJ> = Conjugate used

Search of the numeric version has distinct advantages allowing, for example, range searches. If retrieval of records on monoclonal antibodies reacting with all blood coagulation factors was desired, for example, the range of numerals corresponding to blood coagulation factors in the controlled hierarchical vocabulary list would be entered into the computer. This relieves the searcher of the need to prepare a list of these factors and type each substance name into the computer.

Text-searching, on the other hand, is limited to retrieval of character strings. There is no capacity for grouping of specific search terms. Each term must be entered individually.

Figure 2.1 illustrates a typical output report from a text search.

Fig. 2.1. Example output report from a test search.

```
ACCESSION #          517
AUTHOR/DEVELOPER     Webb KS ; Paulson DF ; Parks SF ; Tuck FL ;
AUTHOR/DEVELOPER     Walther PJ ; Ware JL
ADDRESS              Dept. of Surgery, Division of Urology; Box 3062;
ADDRESS              Duke University Medical Center; Durham, NC 27710 USA
DISTRIBUTOR          Product:Webb KS
DISTRIBUTOR          Dept. of Surgery, Division of Urology; Box 3062;
DISTRIBUTOR          Duke University Medical Center; Durham, NC 27710 USA
REFERENCE            Cancer Immunol Immunother 1984;17(1):7-17
DESIGNATION          Product:alpha-Pro 13 ;developer
IMMUNOGEN             Homo sapiens ;human ;male ;prostate
IMMUNOGEN             carcinoma ;Cell Line:PC-3 ;living
IMMUNOGEN             Homo sapiens ;human ;2.male ;prostate
IMMUNOGEN             carcinoma ;Cell Line:DU145 ;2.living
IMMUNOGEN             Homo sapiens ;human ;3.male ;prostate
IMMUNOGEN             carcinoma ;Cell Line:LNCaP ;3.living
IMMUNIZATION          in vivo ;injection ;multiple antigens ;multiple doses
IMMUNOCYTE DONOR      Mus musculus ;mouse ;Strain BALB/c ;male ;spleen
IMMORTAL PARTNER      Cell Line:Sp2/0
IMMORT.PROCEDURE      fusion
CLASSIF. PRODUCT      IgG2a
ASSAY                 direct binding ;indirect immunoperoxidase
REACTANT      1       Homo sapiens ;human ;prostate ;carcinoma ;Cell Line:PC-
3
REACTANT      1       40 kD ;reduced
REACTANT      2       Homo sapiens ;human ;prostate ;carcinoma ;Cell
Line:DU145
REACTANT      2       40 kD ;reduced
REACTANT      3       Homo sapiens ;human ;prostate ;3.male
REACTANT      3       benign prostatic hypertrophy
CROSSREACTANT1        Homo sapiens ;human ;melanoma
CROSSREACTANT1        40 kD ;denatured
CROSSREACTANT2        Homo sapiens ;human ;blood vessel ;endothelium
CROSSREACTANT3        Homo sapiens ;human ;carcinoma ;kidney
NONREACTANT   1       Homo sapiens ;human ;carcinoma ;prostate
NONREACTANT   1       Cell Line:LNCaP
NONREACTANT   2       Homo sapiens ;human ;liver
NONREACTANT   3       Homo sapiens ;human ;kidney
NONREACTANT   4       Homo sapiens ;human ;spleen
AVAILABILITY         Product:collaborative research only ;supernatant
COMMENTS             Under reducing conditions the antigen has an apparent
COMMENTS             MW of 40 kD with a minor component of 17 kD.
COMMENTS             Under non-reducing conditions the antigen has an
COMMENTS             apparent MW of 120 kD.
COMMENTS             The antigen is relatively stable on the cell membrane;
COMMENTS             neither being shed nor endocytosed readily.
COMMENTS             Antibody reacted with 4/16 clinical samples of prostatic
COMMENTS             cancer by immunoperoxidase technics, 8/10 by solid phase
COMMENTS             binding assays.
```

Specific types of queries which may be answered by searching the HDB are shown below:

Where can monoclonal antibodies against fibronectin be obtained?
Is there a monoclonal antibody for isotype IgM against human hepatoma?
Is there a monoclonal antibody which is specific for chicken neural tissue, i.e. does not show reactivity with neural tissue from any other species?
List the classes of immunoglobulins produced by hybridomas for which the immunising agent was a purified viral antigen.
List the antibacterial products of T-cell clones.
What immunising procedures are used to produce monoclonal antibodies against toxins?
What information can be obtained on the binding sites for monoclonal antibodies directed against cell surface proteins?
Who are the commercial distributors of hybridomas?

As the Data Bank grows and contains a large amount of data on similar monoclonal antibodies produced by different methods and components, comparative studies can be made on immunogens, strain sources, tissue sources and methodology. It is anticipated that this information can be used to improve the technology and provide insight into taxonomic relationships between species, protein structure, antibody binding sites, cell surface receptors and the mechanism of antigen – antibody reactions.

More information on the HDB is available from any of the three nodes:
ATCC
Hybridoma Data Bank
12301 Parklawn Drive
Rockville 20852
USA
Telephone: (301) 231-5585
Electronic mail: BT TYMNET 42:CDT0004

LSRIS
RIKEN
2-1 Hirosawa, Wako
Saitama 351-01
Japan
Telephone: (484) 621111 (ext. 6023)
Telex: 2962818 RIKEN J
Electronic mail: BT TYMNET 42:CDT0007

CERDIC
2° CAI-SOLAREX
Avenue des Maurettes
F-06270 Villeneuve-Loubet
France
Telephone: 93-20-01-80
Electronic mail: TELECOM GOLD 75:DBI0098

Microbial Strain Data Network (MSDN)
The MSDN was started in 1985 in order to construct a world-wide network of holders of strain data serving as Nodes in an informational network. The MSDN will act as a locator service for strains of microbes or cultured cell lines having specific attributes.

From the UNEP-initiated round table in 1982 (at the International Congress for Microbiology), where the concept was introduced, to the present time, the Network has been designed. An infrastructure has been established, consisting of a policy-making Task Group, six operating Committees, and a Secretariat; and pilot on-line databases have been installed on a publicly accessible commercial computer host. Sponsorship is by three components of the International Council of Scientific Unions: CODATA, WFCC, and IUMS.

The sheer magnitude of the number of repositories of microbial strain data and the numbers of strains held within those repositories generate serious data acquisition and communication problems. Collecting the desired data in one or a few places for general availability is a practical impossibility. Rather, the MSDN is designed to operate as a locator service for repositories of desired combinations of attributes by an indirect method. Thus, the data repositories become Network Informational Nodes.

The Central Directory of the MSDN contains a list of the data elements recorded by all the various Nodes rather than the data themselves. This database is a controlled vocabulary of standardised nomenclature of mainly biochemical and morphological features used for characterisation throughout the world. The initial basis for the controlled vocabulary comes from the CODATA-sponsored publication by Rogosa, Krichevsky & Colwell (1986). The user scans the vocabulary to select features of interest. The features in the form of the controlled terms are entered as search criteria into a second database which yields the names of repositories assessing the strains for possession of the desired features. When a Node is located which codes information on the desired data elements (features), the person querying the MSDN contacts the Node(s) directly for existing clones fitting the detailed patterns of attributes. The contact may be accomplished through tele-communications (where the capabilities exist) or by mail or telephone. In an increasing number of cases, direct access to host computer databases is possible.

In addition to the Central Directory, a number of other databases and services are available from the MSDN (see Tables 2.2 and 2.3). Thus, the HDB database provides a simple mechanism for locating hybridomas and/or monoclonal antibodies that their distributors state are available to the scientific community. Again, a number of culture collection catalogues are on-line with frequent updates. On-line ordering of cultures or requests for supplementary information on cultures is available from the culture collections maintaining a sales mailbox or catalogue on the system.

Gateways to various collection computers have been established or

Table 2.2 *MSDN Services*

MSDN Central Directory
Other databases (HDB, MiCIS, collection catalogues. . .) either stored on MSDN computer or made available through electronic gateway
On-line culture ordering facility
Electronic mail (including Telex and Fax)
Bulletin Board
Computer conferences
Micro-IS software distribution
Training
Access to BT TYMNET and TELECOM GOLD services (e.g. International Air Line Guide)
Access to people on other BT TYMNET/TELECOM GOLD systems
User support

are under development. The first one in place was to the National Collection of Yeast Cultures (UK) (NCYC) computer. This gateway provides direct access to the catalogue as well as a yeast identification system. Other gateways are to the World Data Center for Collections of Microorganisms (WDC) in Japan, the MiCIS database, the Centraalbureau voor Schimmelcultures database and DATA-STAR databases. General gateways are available to the 'academic networks'.

The electronic mail service on the MSDN system is a powerful communication tool in its own right. The system has integrated bulletin board and computer conferencing facilities. In addition, the mail system is linked to the telex, cable and fax systems.

Services of the MSDN that are not electronic include organising and conducting training courses in the use of computers in microbiology and international electronic communication as well as distribution of the software of the Microbial Information System (MICRO-IS) on a shareware basis.

Queries to the MSDN may be through mail, telephone, or through the BT TYMNET and TELECOM GOLD services which link to most

Table 2.3. *Databases available through the MSDN network*

MSDN Central Directory – locates centres with specific information on the properties of microbial and cultured cells. A 'yellow pages' Directory referring users to sources of primary data. Data strictly defined in scientific terms, using a numeric coding system. World-wide data providers.
Hybridoma Data Bank – contains data on publicly available immunoclones and their products (USA, Europe, Japan nodes)
Information Centre for European Culture Collections (MiCIS and DSM databases)[a]
Databases of the Centraalbureau voor Schimmelcultures, Baarn, Netherlands[a]
World Data Center on Collections of Microorganisms, Japan[a]
DATA-STAR databases[a]
ATCC Recombinant clones and libraries
NCYC Computer Services[a]
Culture collection catalogues:
American Type Culture Collection animal cells
American Type Culture Collection algae and protozoa
American Type Culture Collection bacteria
CAB International Mycological Institute
European Collection of Animal Cell Cultures
UK National Collection of Food Bacteria (NCFB)[a]
UK National Collection of Yeasts (NCYC)[a]

[a] Accessed via electronic gateway.

common packet switching services as well as to the telex and TWX systems.

Contact:
MSDN Secretariat
Institute of Biotechnology
Cambridge University
307 Huntingdon Road
Cambridge CB3 0JX
UK
Telephone: (0223) 276622
Telex: 812240 CAMSPL G
Fax: (0223) 277605
Electronic mail: TELECOM GOLD 75:DBI0001/DBI0005
Janet: MSDN@ PHX.CAM.AC.UK

World Data Center for Collections of Microorganisms (WDC)
In 1984, the World Federation for Culture Collections (WFCC) officially reaffirmed the WDC as a component of the WFCC and thus accepted responsibility for the operation and management of the WDC. Because of the announcement of the retirement of the director of the WDC at Brisbane, Australia, a public search was conducted for potential hosts to ensure continuity of the WDC effort. After competitive evaluation of the proposals, the Executive Board of the WFCC agreed that the WDC be transferred to the Institute of Physical and Chemical Research (RIKEN), Saitama, Japan.

WDC is an information centre which supports culture collections and their users. The primary tasks are:

(1) publication of descriptions of culture collections;
(2) production of a species-oriented directory of culture collection holdings.

The WDC publishing plans for 1989 included a World Directory of Algae (completed), a World Directory of Protozoa, and a Directory of Asian culture collections.

The tasks of WDC are not limited to the above; other tasks will be performed based on necessity and available resources.

Information sources for WDC are culture collections and national/local/international data centres as well. The WDC makes use of, and cooperates with, HDB and MSDN.

For information dissemination, the WDC uses a variety of communication media such as mail, telex, cable, fax, electronic mail, publication and magnetic devices (floppy disks and magnetic tapes).

The WDC currently holds information on 327 culture collections distributed over 56 countries. The core data of the WDC are descriptions of culture collections and their holdings allowing users to locate a culture collection and/or taxonomic category of microorganism. They can find strains of specific species by consulting the list of species preserved in culture collections. The WDC database currently includes bacteria, fungi, yeasts, algae, protozoa, lichens, animal and plant cells, and viruses.

The WDC participated in a survey and workshop on an 'Asian Information Network on Biomaterials' and is cooperating in a feasibility study of such a network.

The WDC is able to answer queries by any communication media including on-line retrieval.

Contact:
WDC/RIKEN
2-1 Hirosawa, Wako
Saitama 351-01
Japan
Telephone: (484) 621111
Telex: 2962818 RIKEN J
Electronic mail: BT TYMNET 42:CDT0007

World Federation for Culture Collections (WFCC)

The World Federation for Culture Collections provides information to biotechnologists in a variety of ways within the overall mission of the WFCC as described in Chapter 8. Specific information initiatives described above are the World Data Center for Collections of Microorganisms and the Microbial Strain Data Network. Specialist committees established by the WFCC can provide information on a number of topics. The Patents Committee considers patent conventions. The Postal and Quarantine Committee is a good source of information on regulations governing shipment of cultures. The Publicity Committee publishes a newsletter on a periodic basis. The Committee on Endangered Collections identifies and takes action to rescue jeopardised collections. The Education Committee initiates training activities (e.g. books, videotapes, courses, individual instruction).

Contact:
World Federation for Culture Collections
Institute of Biotechnology
Cambridge University
307 Huntingdon Road
Cambridge CB3 0JX
UK
Telephone: (0223) 276622
Telex: 812240 CAMSPL G
Fax: (0223) 277605
Electronic mail: TELECOM GOLD 75:DBI0005
Janet: MSDN@PHX.CAM.AC.UK

Directory of Biotechnology Information Resources (DBIR)
There is a variety of newsletters, bulletin boards, and other information sources that cover various aspects of biotechnology in addition to those listed above. Finding these can be difficult in many instances. An initiative of the Specialized Information Services Division of the National Library of Medicine in the USA, in collaboration with the Bioinformatics Department of the American Type Culture Collection, has the purpose of helping with this problem.

The Directory of Biotechnology Information Resources (DBIR) is a centralised directory to international sources of publicly available biotechnology information such as: computerised databases and their distributors, networks, electronic bulletin boards, and other biological computer resources established for communicating and disseminating biotechnology data; culture collections and specimen banks; biotechnology centres and other organisations which stimulate biotechnology growth in academia and industry; publications focusing on general issues in biotechnology, which include selected directories, serials, monographs, reviews, and compilations; and nomenclature committees working on issues of nomenclature in biotechnology and molecular biology.

The Directory encompasses a broad spectrum of biotechnology, including its application in medicine, agriculture, pharmaceuticals, food technology, and microbiology, as well as molecular biology. Because of the international and interdisciplinary scope of the Directory, it should prove especially useful to those requiring an entry point into unfamiliar areas of biotechnology.

DBIR is currently maintained by MEDLARS on the National Library of Medicine's (NLM) Toxicology Data Network (TOXNET) and as a part of the Directory of Information Resources Online (DIRLINE) file.

Contact:
Directory of Biotechnology Information/Resources
Bioinformatics Department
American Type Culture Collection
12301 Parklawn Drive
Rockville
Maryland 20852-1776
USA
Telephone: (301) 231-5585
Electronic mail: DIALCOM 42:CDT0004

or Specialized Information Services Division
National Library of Medicine
8600 Rockville Pike
Bethesda
Maryland 20894
USA
Telephone: (301) 496-6531 or (301) 496-1131

2.4 Access to data resources

2.4.1 *Traditional*

Traditional methods for accessing data resources are still the most common and are quite satisfactory as long as the answers are not voluminous and are not needed quickly. Asking a culture collection curator about the availability of a single strain exemplifying a particular taxon or having a particular set of attributes will usually get a prompt reply. The query may be verbal, in person or on the telephone, or by mail. As the queries become more complex, the use of these methods become less satisfactory to both the seekers and providers of information.

The seeker of information becomes frustrated at the delay and incomplete nature of the answer that comes back. Often multiple cycles are required to refine the question to the point where the desired answer is given. While this refinement of communication is valuable in clarifying the true nature of the query (not always recognised from the initial inquiry) and common to all pathways, the frustration is amplified by the length of the cycle time.

The provider of information must devote increasing resources to answering queries. This takes professional expertise that could be used in the other work of the collection. Therefore, any mechanism which minimises the labour involved in answering queries increases the professional resources of the collection.

Both internal and external mechanisms are useful to alleviate some of the workload. Internal mechanisms include publications, such as the aforementioned catalogues, and computer management of the data for ease of searching and reporting. External mechanisms are primarily electronic forms of communication and are discussed in the next section.

2.4.2 *Electronic*

Except for voice communication via telephone, telecommunication has only recently become a part of collection life. Many of the larger collections have been using printed electronic messages (cables, telex) for a while. However, the advent of computer-operated message transfer systems at reasonable cost (e.g. direct on-line access, electronic mail systems, and public packet switching services) have allowed economically practical electronic communication between questioner and answering resource.

Two parallel paths of development are taking place at this time in providing public access to data in collections. Some collections are following both paths simultaneously.

In the first instance, a collection may make its data accessible to the public by establishing access to a computer system maintained by the collection or its parent institution. Some examples are the Human Gene Probe Bank at the American Type Culture Collection, the National Collection of Yeast Cultures at Norwich, UK, the Japan Collection of Microorganisms, Tokyo, and the CAB International Mycological Institute, Kew, UK.

The second approach is to install the data in a computer operated by others. The data and their installation may be accomplished under the control of the collection as is being done by the ATCC on the MSDN CODATA Network. Alternatively, all or part of the data may be installed on a computer operated as a national or regional facility such as those described above for Europe, Brazil, and Japan (MiCIS, MINE, Catalogo Nacional De Linhagens, and NISLO).

In all these cases of providing access through electronic computer services, the source of the data does not maintain the communication paths beyond the host computer. It is generally the responsibility of the seeker of information to find the most appropriate path. Unless the provider and seeker of information are at the same institution, where direct connection to the computer may be possible, the ordinary telephone system is likely to be the first resource connected to the seeker's terminal. Where short distances are involved, it may be reasonable to use only the voice-carrying telephone system. However, national and international data communications are using 'packet switching service' (PSS) for an ever-increasing share of data telecommunications. Such PSS transmission is cheaper and more reliable by far than direct telephone calls.

To find out more about telecommunications with the systems described in this chapter, get in touch with the appropriate system listed. Help with establishing electronic communication paths may be obtained from the MSDN (see above).

2.5 References

Bussard, A., Krichevsky, M. I. & Blaine, L. D. (1985). Hybridoma Data Bank: structure and function. In *Monoclonal Antibodies Against Bacteria*, vol. I, ed. A. J. Macario & E. Conway de Macario, pp. 287–311. Orlando: Academic Press.

Rogosa, M., Krichevsky, M. I. & Colwell, R. R. (1986). *Coding Microbiological Data for Computers*. New York: Springer-Verlag.

3

Administration and safety

A. DOYLE and K. ALLNER

3.1 Supply of cultures

The administrative procedures for obtaining cultures vary from collection to collection. As well as postal enquiries, most collections also will accept orders by telephone, telex, fax or electronic mail.

Information on charges and other administrative details may be found in published catalogues, and moves are currently under way to establish electronic catalogues that may be searched on-line; both the American Type Culture Collections (ATCC) and European Collection of Animal Cell Cultures (ECACC) are available on-line via the Microbial Strain Data Network (MSDN). Although information by these routes is valuable, it does not replace the extensive knowledge and experience of the staff of the collections regarding the deposits held. Sources of data are further described in Chapter 2.

3.1.1 *Deposit restrictions*

The major aim of culture collections is to make cultures readily available to the scientific community. However, contractual obligations dictate that in some cases restrictions are placed on distribution, and forms of agreement have to be signed prior to the release of certain cultures. The categories are:

(1) Cell lines available without restriction. In the case of ATCC this applies to Certified Cell Lines (CCL) and Cell Repository Lines (CRL).

(2) Restricted distribution cell lines. An agreement form is used by the ATCC to cover the Hybridoma Bank and Tumour Immunology Bank deposits. Different types of forms exist for government (Fig. 3.1) and non-government (Fig. 3.2)

Fig. 3.1. ATCC form of agreement: Government.

GOVERNMENT INSTITUTIONS
FORM TO BE COMPLETED BEFORE ATCC CAN SHIP CERTAIN CELLS

Cells Provided From The Human Tumor Cell Bank and/or The Hybridoma Bank (under NIH Contracts) and some of the TIB lines to Investigators at Government Institutions

I hereby agree that the cell lines provided are for research purposes only. Cell lines and their products shall not be sold or used for commercial purposes. Nor will cells be distributed further to third parties for purposes of sale, or producing for sale, cells or their products. Secondary distribution shall only be made under the terms outlined in this form. The cells are provided as a service to the research community. They are provided without warranty or merchantability of fitness for a particular purpose or any other warranty, express or implied.

Accepted:

Typed or Printed Name ____________ Agency ____________

Signature ____________ Date ____________ Division or Department ____________

City/State ____________

Cell Lines of Human Origin

I understand that, although human cells distributed by the American Type Culture Collection have been subjected to stringent tests and observations which indicate the absence of extraneous agents and deleterious properties, the ATCC accepts no responsibility for any injury (including injury resulting in death), damage or loss that may arise from the use of the cells, either directly (including use for diagnostic purposes) or in the preparation of a product. I assume all risks and responsibility in connection with their receipt, handling, storage and use.

Typed or Printed Name ____________ Agency ____________

Signature ____________ Date ____________ Division or Department ____________

City/State ____________

Upon receipt of this signed understanding, the ATCC will be able to meet this request for these cells and any further requests you may make in the future.

Human Tumor Cell Bank, brief statement of intended use:

____

____

____

American Type Culture Collection
12301 Parklawn Drive
Rockville, MD 20852 USA

telex: 908768 ATCCROVE
FAX: 301-231-5826

FORM-100-G

Fig. 3.2. ATCC form of agreement: non-Government.

NON-GOVERNMENT INSTITUTIONS
FORM TO BE COMPLETED BEFORE ATCC CAN SHIP CERTAIN CELLS

Cells Provided From The Human Tumor Cell Bank and/or The Hybridoma Bank (under NIH Contracts) and some of the TIB lines to Investigators at Non-Government Institutions

I hereby agree that the cell lines provided are for research purposes only. Cell lines and their products shall not be sold or used for commercial purposes. Nor will cells be distributed further to third parties for purposes of sale, or producing for sale, cells or their products. Secondary distribution shall only be made under the terms outlined in this form.

The cells are provided as a service to the research community. They are provided without warranty or merchantability of fitness for a particular purpose or any other warranty, express or implied. In addition, the recipients of the cell lines agree to indemnify and hold harmless the United States from any claims, costs, damages, or expenses resulting from any injury (including death), damage, or loss that may arise from the use of the cell lines.

Accepted:

Typed or Printed Name　　Institution

Signature　　Date　　Department

City/State

Cell Lines of Human Origin

I understand that, although human cells distributed by the American Type Culture Collection have been subjected to stringent tests and observations which indicate the absence of extraneous agents and deleterious properties, the ATCC accepts no responsibility for any injury (including injury resulting in death), damage or loss that may arise from the use of the cells, either directly (including use for diagnostic purposes) or in the preparation of a product. I assume all risks and responsibility in connection with their receipt, handling, storage and use.

Typed or Printed Name　　Institution

Signature　　Date　　Department

City/State

Upon receipt of this signed understanding, the ATCC will be able to meet this request for these cells and any further requests you may make in the future.

Human Tumor Cell Bank, brief statement of intended use:

American Type Culture Collection
12301 Parklawn Drive
Rockville, MD 20852 USA

telex: 908768 ATCCROVE
FAX: 301-231-5826

FORM-100-NG

institutions. Release of deposits is conditional on this form being received. This is also the case for certain cell lines held by the ECACC (Fig. 3.3).

(3) Liability waiver. In addition to the type of release conditions given above, human cell lines are released by the ATCC or the Corriell Institute for Medical Research (CIMR) on condition that an additional agreement concerning liability during use is signed. An example of the CIMR form is shown in Fig. 3.4.

3.1.2 *Prices*

Service culture collections are required to offset some of their running costs by charging for the supply of cultures. Differential systems are in operation in some cases; the ATCC, for example, has a US and Canadian domestic rate for cell lines and a higher rate for commercial and foreign institutions. By contrast CIMR lines are charged at a differential rate for commercial and non-profit institutions whether in the US or abroad. ECACC operates the same price structure for commercial and non-profit organisations both at home and over-

Fig. 3.3. ECACC form of agreement.

AGREEMENT BETWEEN THE EUROPEAN COLLECTION OF ANIMAL CELL CULTURES AND RECIPIENT OF CELL LINE(S)

The cell line(s) listed below will be made available subject to the following conditions.

1. The cultures are to be used either by
   a) Scientists affiliated to a non-profit Research Institute in pursuance of an academic research programme.
   or
   b) if by commercial or similar interests, they will not be offered for sale or be utilised in commercial processes involved in the preparation for sale of any biological material or in any other type of commercial activity without having secured written agreement from the ECACC on behalf of the original depositor. ECACC will refer back to depositors all such requests.
2. The cell line(s) will not be distributed to third parties.
3. If the cell line(s) are referred to in any publication, then correct reference will be made to the work of the original depositor and no alteration will be made to its ECACC title or acronym.
4. It is understood that neither ECACC or its depositors accept any liability whatsoever in connection with the handling or use of the cell line(s).

Signed for and on behalf of ....

By the recipient ....

Name: .... Date: ....

Position: ....

Address: ....

....

Cell line(s) ....

....

EUROPEAN COLLECTION OF ANIMAL CELL CULTURES

PHLS Centre for Applied Microbiology and Research, Porton Down, Salisbury, Wilts. SP4 0JG, UK

seas. However, certain lines from ECACC are available free of charge to non-profit organisations in Europe due to a grant funding agreement with the UK Research Councils.

In all cases there are additional shipping charges dependent upon the mode of shipment. If air-freight is used the additional costs can be significant.

3.1.3 *Shipping*

Animal cells can be shipped, either as growing cultures or as frozen cultures in ampoules and there are advantages and disadvantages to both. Growing cultures will survive longer in transit provided that great fluctuations in temperature are not expected, but will need immediate handling on receipt. An additional charge may be made for

Fig. 3.4. CIMR compliance form.

CIMR COMPLIANCE FORM

Waiver and Prior Agreement Before Receipt of Repository Cell Culture #

"Ownership of cell line # in the repository resides with the original developer. No distribution of the cell line shall be made prior to signed agreement of this waiver. The cell line is provided for research purposes only, and it or its products shall not be sold or used for commercial purposes, nor will cells be distributed further to third parties for purposes of sale, or producing for sale cells or their products. Secondary distribution shall only be after written approval by the contributor of the cell line. The cells are provided as a service to the research community. They are provided without warranty of merchantability or fitness for a particular purpose or any other warranty, expressed or implied. In addition, the recipients of the cell lines agree to indemnify and hold harmless the United States government and CIMR and the contributor from any claims, costs, damages, or expenses resulting from any injury (including death), damage, or loss that may arise from the use of the cell line."

Date ____________ Signed ____________
Name of Investigator

____
Institution

Signed ____________
Department Chairman

the dispatch of a growing culture. If there are particularly unusual medium requirements for the cell line this can lead to difficulties. On the other hand, frozen cells are shipped in dry ice (−70 °C) and have a shorter life expectancy in transit, but once received they may be returned to liquid nitrogen storage prior to resuscitation if the medium necessary for growth is not immediately available. However, it is normally recommended that cells are thawed on receipt as viability problems can result from re-freezing to −196 °C.

3.1.4 *Air-freight*

There are many international air-freight companies experienced in the shipment of perishable cargo and most collections use their services for shipment overseas. Prior to dispatch the customer is informed, by telex or telephone of the date of arrival, flight number and airway bill number of the consignment. Customs clearance and delivery from the airport are usually the responsibility of the customer. These formalities can also be dealt with by a handling agent if necessary. However, certain services operate a door-to-door delivery service, but certain local taxes may be payable on receipt.

3.1.5 *Postal services*

A live culture can be expected to survive up to five days in transit at ambient temperature. Therefore dispatch by an express postal service is acceptable, and is also a considerably cheaper option than air freight or courier services. For overseas, each dispatch requires a completed customs declaration to be attached.

3.1.6 *Import/export requirements*

There are certain national regulations regarding the importation of cell lines into a country. For example, there are currently no import restrictions in force in the United Kingdom; however, in the case of the USA there is concern over the nature of the growth media and other reagents in cell culture. Concern over the possible importation of material carrying the foot and mouth disease virus means that material entering the USA has to be approved either by *in vivo* or *in vitro* testing. In some cases the provision of certificates of origin for bovine serum and trypsin, the two particular reagents in question, can speed matters. It is essential, that these should be derived from acceptable sources and that the countries of origin of these reagents should be free from foot and mouth disease virus infection. It is always advisable to check the national regulations before embarking on the import or export of cell lines.

3.2 Health and safety

Although cell lines distributed by culture collections have undergone characterisation and quality control (most rigorously applied in the case of certified cell lines) consideration must be given to the limitations of the techniques used to examine the cells. If, to take an extreme example, a cell line is patient-derived, during the early stages of culture the environmental conditions may not have provoked overt appearance of a contaminating virus. This could later be induced from the cells and if the material is handled in an uncontained environment possibly lead to infection of laboratory staff. Culture collections have to be aware of their obligations as suppliers of cell cultures and legislation in the UK – under the Health and Safety at Work etc. Act 1974 and the Control of Substances Hazardous to Health Regulations 1988 – means that there are certain legal implications to the supply of cultures for use in laboratories. The recommendation that ECACC currently gives on the use of uncharacterised material is to handle the cultures under conditions which offer both operator and cell culture protection.

Fortunately, occurrences of laboratory acquired infection from cell cultures considered to be free of potential pathogens are rare. However, a cautious approach to the handling of cell lines that are not fully characterised is good and sensible laboratory practice.

3.2.1 *Laboratory practice*

Since the introduction of the Occupational Safety & Health Act 1970 in the USA and the Health and Safety at Work etc. Act in 1974 in the UK, a spate of guidelines, regulations or recommendations has been produced pertaining to the handling of microorganisms in research and production establishments, hospitals and educational departments. There has been a conscious attempt to ensure that acceptable safety procedures are established and maintained at all places of work. In the USA in 1984, the Centers for Disease Control and National Institutes of Health have jointly prepared *Biosafety in Microbiological and Biomedical Laboratories*, which describes combinations of standard and special microbiological practices, safety equipment and facilities that constitute Biosafety Levels 1–4 recommended for working with a variety of infectious agents. The four classes and examples of organisms in each class are listed in Table 3.1. Table 3.2 outlines recommended laboratory handling procedures for infectious agents.

In the UK the Health and Safety Executive offers an advisory service to the community in all matters of safety and has inspectoral powers to enforce the law in instances where the Health and Safety at Work Act is breached. The latest guidance on the categorisation of pathogens is to be found in a report produced by the Advisory Committee on Dangerous Pathogens (ACDP), 1984, *Categorisation of Pathogens according to Hazard and Categories of Containment*. Bacteria, Chlamydia, Rickettsiae, mycoplasmas, fungi, viruses and parasites are clearly categorised according to the hazard they present to workers and the community, and four hazard groups are identified (1–4). Information is given on the degree of containment and protective clothing which should be

Table 3.1. *Classification of microorganisms according to biological hazard and their shipping requirements*[a]

Class I	Agents of no recognised hazard under ordinary conditions
Examples	*Saccharomyces cerevisiae, Trichoderma reesei, Lactobacillus casei*
Shipment	Culture-tube in fiberboard or other container. Permits as required
Class II	Agents of ordinary potential hazard
Examples	*Aspergillus fumigatus, Candida albicans, Cryptococcus neoformans, Staphylococcus aureus*
Shipment	Culture-tube wrapped in absorbent material, placed in metal screw-cap can, placed in fiberboard container. Permits as required
Class III	Pathogens involving special hazard
Examples	*Coccidioides immitis, Ajellomyces capsulatum, Bacillus anthracis, Yersinia pestis*
Shipment	Culture-tube heat sealed in plastic, wrapped in absorbent material, placed in hermetically sealed can, placed in sturdy cardboard box. Permits as required. Etiologic agent warning label necessary
Class IV	Pathogens of extreme hazard
Examples	*Arthroderma simii, Pasteurella multocoida,* certain animal-/plant viruses
Shipment	Culture-tube heat sealed in plastic, wrapped in absorbent material, placed in hermetically sealed can, placed in sturdy cardboard box. Required permits. Etiologic agent warning label necessary

[a] US Department of Health, Education and Welfare, 1972; US Department of Health and Human Services, Public Health Service, 1983.

applied during the handling of such organisms in the laboratory, including requirements for animal containment.
Examples are:

Epstein–Barr Virus Hazard Group 2, e.g. B95–8 marmost cell line

Hepatitis B Virus Hazard Group 3, e.g. Hep 3B cell line, secretes Hepatitis B surface antigen

Table 3.2. *Summary of recommended biosafety levels for infecting agents*[a]

Biosafety level	Practices and techniques	Safety equipment	Facilities
1	Standard microbiological practices	None: primary containment provided by adherence to standard laboratory practices during open-bench operations	Basic
2	Level 1 practices plus: laboratory coats; decontamination of all infectious wastes; limited access; protective gloves and biohazard warning signs as indicated	Partial containment equipment (i.e. Class I or Class II Biological Safety Cabinets) used to conduct mechanical and manipulative procedures that have high aerosol potential that may increase the risk of exposure to personnel	Basic
3	Level 2 practices plus: special laboratory clothing; controlled access	Partial containment equipment used for all manipulations of infectious material	Containment
4	Level 3 practices plus: entrance Maximum containment through change room where street clothing donned; shower on exit; all wastes decontaminated on exit from facility	Maximum containment equipment (i.e. Class III biological safety cabinet or partial containment equipment in combination with full-body, air-supplied positive-pressure personnel suit) used for all procedures and activities	Maximum containment

[a] US Public Health Service, 1983.

Korean Haemorrhagic Fever Hazard Group 4
Lassa Fever Virus Hazard Group 4

The American and British guidelines vary slightly but both have been accepted by the World Health Organisation. Other countries have developed similar systems or use those operating in the USA or the UK.

The control of genetic manipulation experiments in the UK is the responsibility of the Advisory Committee on Genetic Manipulation, using its own classification of experiments 1–4. Guidance is offered in a series of newsletters which are constantly updated.

The American control system is somewhat more complicated but all Federal agencies that fund research related to biotechnology adhere to the policy that research in this field must conform to the requirements of the coordinated framework, such as the National Institutes of Health Recombinant DNA Guideline. Other countries engaged in recombinant DNA techniques have local guidelines or use the American or British guidelines.

3.2.2 *Supply of cultures*

Culture collections are aware of the responsibilities associated with supply of cultures and require *bona fide* signatories before releasing hazardous pathogens. The National Collection of Type Cultures (NCTC) in the UK issues a leaflet to prospective recipients, explaining that all cultures supplied by them must be regarded as potentially pathogenic and be handled by or under the supervision of competent persons trained in microbiological techniques. This includes compliance with national or local codes of practice.

In the USA the American Type Culture Collection (ATCC) requires evidence that the recipient is trained in microbiology and has access to a properly equipped laboratory with the appropriate containment facilities. Requests for Class III pathogens must be accompanied by a signed statement assuming all risks and responsibility for subsequent use.

Customers ordering and receiving cultures from the German collection of microorganisms (Deutsche Sammlung von Mikroorganismen, DSM) assume all responsibility on receipt for the manner in which the cultures are handled and stored. The DSM does not accept liability for any injuries arising from the supply of cultures. Scientists within the German Federal Republic should consult the *Vorläufige Empfehlungen für die Klassifikation von Mikroorganismen und Krankheitserregern nach den im Umgang mit ihnen auftretenden Gefahren* for details.

3.2.3 *Transportation of cultures*

The distribution of cultures, both domestically and overseas, is essential to further scientific endeavour world-wide. However, the shipment of microbial cultures subjects culture collections, as well as individual scientists, to a complex set of laws that have been designed to protect humans, plants and animals from disaster through accidental release or introduction of infectious agents. Most countries have laws concerning both the import and the export of microbial cultures, and frequently the states and provinces of these countries also regulate movement of pathogenic materials. Accordingly, those who ship or import cultures need to ensure compliance with these laws and regulations.

With regard to transport by air, the International Civil Aviation Organisation (ICAO) and International Air Transport Association (IATA) have reached agreement with curators of culture collections on procedures to be followed. Information is published by ICAO (1985) providing details of the designation of dangerous goods, with packing instructions, which may be sent by air. A list of air operators holding general permission to carry such goods by air is also available. This information is continually updated and reviewed.

The postal authorites within the UK will also accept cultures for transportation either domestically or overseas. Conditions for the acceptance of Infectious Perishable Biological Substances (IPBS) must be sought from the Post Office Authorities prior to dispatch, usually from a nominated Crown Post Office. Dispatch of perishable biological substances in the overseas post is restricted to those countries whose postal administrations are prepared to admit such items. A list of countries accepting these items appears in the Post Office Guide, available from certain HMSO Offices (see Section 3.3, Further reading). Senders are strongly recommended to ascertain from the addressee before dispatch that the goods will be acceptable in the country of destination. Infectious biological materials will be accepted by British Rail on a station-to-station basis (i.e. no collection service) by either a Recorded Parcels transit scheme or Red Star Premium Service. Terms and conditions using the United National Classification can be found in British Rail's publication (see Further reading).

In the UK the information on legislation governing the import and export of infectious biological substances can be obtained from either the Health and Safety Executive (HSE), Magdalen House, Stanley Precinct, Bootle, Merseyside, or the Ministry of Agriculture, Fisheries and Food (MAFF), Hook Rise South, Tolworth, Surbiton, Surrey.

In the USA the importation of infectious biological substances capable of causing human disease is controlled by the Public Health Service Foreign Quarantine Regulations (42, Code of Federal Regulations, Section 71.156). Importation permits, conditions of shipment and handling procedures are issued by the Centers for Disease Control. Similar restrictions and conditions for the importation of animal pathogens other than those affecting humans are subject to US Department of Agriculture regulations, Hyattsville, Maryland. The ATCC (1986) provides a summary of regulations affecting US scientists. The National Institutes of Health (NIH) have published a Laboratory Safety Monograph which provides additional information.

There is no doubt that continued interest and concern will be expressed by both the scientific community and the general public about the safety aspects surrounding issues pertaining to microbiological agents. Safety professionals and scientists alike have an obligation to ensure that all microbiological agents are handled in a safe and responsible manner. It is impossible to detail procedures for all countries here but reference to the following reading list should enable the reader to obtain the necessary information. Information on appropriate packing material and suppliers can be obtained through the service culture collections listed in Chapter 1. A number of culture collections, such as the ATCC, provide in their catalogues useful information on aspects of safety or have advisory leaflets available for distribution.

3.3 Further reading (in chronological order)

The Occupational Safety and Health Act of 1970. United States Occupational Safety and Health Administration. Public Law 91-596. 91st Congress S.2193. USA.

US Department of Health, Education and Welfare (1972). *Classification of Etiologic Agents on the Basis of Hazard*. Centers for Disease Control, Atlanta, Georgia 30333, USA.

The Health and Safety at Work, etc. Act 1974. HMSO, London.

List of Dangerous Goods and Conditions of Acceptance by Freight Train and by Passenger Train, BR. 22426 (1977, under revision). Available from Claims Manager, British Rail Board, Marylebone Passenger Station, London NW1 6JR, UK.

Laboratory Safety Monograph: A Supplement to NIH Guidelines for Recombinant

DNA Research (1978). National Institutes of Health, Bethesda, Maryland, USA.

Vorläufige Empfehlungen für den Umgang mit pathogenen Mikroorganismen und für die Klassifikation von Mikroorganismen und Krankheitserregern nach dem im Umgang mit ihnen auftretenden Gefahren. Published in *Bundesgesundheitsblatt* **23**, 347–59 (1981).

US Department of Health and Human Services, Public Health Service (1983). *Biosafety in Microbiological and Biomedical Laboratories*. National Institutes of Health, Bethesda, Maryland, USA.

Categorisation of Pathogens according to Hazard and Categories of Containment (1984). Advisory Committee on Dangerous Pathogens. HMSO, London. ISBN 0-11-883761-3.

Biosafety in Microbiological and Biomedical Laboratories (1984). US Department of Health and Human Services, Public Health Services, Centers for Disease Control and National Institutes of Health, HHS publication No. (CDC) 84-8395.

The Air Navigation (Dangerous Goods) Regulations (1985). English language edition of the International Civil Aviation Organisation. Technical Instructions for the Safe Transport of Dangerous Goods by Air (DOC 9284-AN/905).

American Type Culture Collection (1986). *Packaging and Shipping of Biological Materials at ATCC*. Rockville, Maryland, USA.

LAV/HTLV III – the causative agent of AIDS and related conditions. Revised guidelines (1986). Advisory Committee on Dangerous Pathogens, DHSS. Health Publications Unit No. 2 Site, Manchester Road, Heywood, Lancs. OL10 2PZ, UK.

Undated references

Advisory Committee on Genetic Manipulation. Guidelines and newsletters. ACGM Secretariat, Baynards House, 1 Chepstow Place, London W2 4TF, UK.

Conditions of Supply of NCTC Cultures: Hazardous Pathogens. Public Health Laboratory Service, Central Public Health Laboratory, 61 Colindale Avenue, London NW9 5HT, UK.

Importation Permits available from Centers for Disease Control, 1600 Clifton Road, NE, Atlanta, Georgia, USA.

Public Health Service Foreign Quarantine Regulations, 42 CFR. Section 71-156.

The Post Office Guide. Available from certain HMSO Offices, including 49 High Holborn, London WC1V 6HB, UK.

4

Culture and maintenance

A. DOYLE, J. B. GRIFFITHS, C. B. MORRIS and D. G. NEWELL

4.1 Routine culture and growth media

Although many books have been written on this subject, certain basic principles of technique can be outlined here by way of introduction to the topic. The majority of animal cell lines falls into adherent or non-adherent categories and the type of cell line influences the choice of growth medium. These vary from complex medium to even more complex medium (Minimum Essential Medium MEM to RPMI 1640), and have a well defined list of ingredients with known buffering capacity. They are supplied in either powder form for preparation and sterilisation in the laboratory, or in bulk liquid form in anything from 1× to 10× strength.

The choice of medium depends on the circumstances of the laboratory. The preparation of powdered medium and its sterilisation is time-consuming and labour-intensive and many laboratories would prefer to leave its preparation and the task of quality control to a commercial supplier. However, since other components of liquid media (e.g. L-glutamine, serum) will have to be added prior to use, it is recommended to carry out a programme of quality control for the microbiological purity of media and for the most expensive constituent of media, the serum, in the laboratory. Most hybridomas are fastidious in their medium requirements and the Foetal Bovine Serum (FBS) should be pre-screened for its ability to maintain hybridoma growth and cloning efficiency. As hybridomas present particular problems a separate section is devoted to their maintenance (4.2.4).

Another consideration in the purchase of serum is the country of origin. Certain areas of the world are still considered to be endemic for Foot and Mouth Disease Virus. Using serum from any country of origin other than 'zone 1' (USA, Canada, UK and New Zealand) may well lead to difficulties in importing cell lines into the United States. If this kind of consideration is important, some caution must also be advised in the source of trypsin used in the passage (the term given to removal of cell lines from one vessel to another) of adherent cell lines. Some details regarding routine subculture are given in the following pages. Further details are available in the recommended texts.

4.2 Storage and maintenance of cell cultures

The fact that the number of cell lines available to the research worker today can be counted in thousands is attributable to the discovery that animal cells can be stored for long periods at sub-zero temperatures. This allows a research worker to store sufficient material for future use after a project is finished, and avoids the need for the almost impossible burden of continually maintaining the culture by serial passage.

Cryopreservation is the term given to the method of storing cells. For optimum results, storage in liquid nitrogen at a temperature of −196 °C is used, as a temperature of below −130 °C is thought necessary to maintain cell viability and stability over a period of years. For short-term storage, a temperature of −80 °C is sufficient to maintain viability for a few months, but usually no longer.

The basic requirements for cryopreservation applied to animal cells can be summarised as follows:

- slow freeze (−1 ° to −3 °/min^{-1})
- a cryoprotectant is necessary; glycerol or dimethyl sulphoxide (DMSO) is suitable
- high protein concentration (provided by bovine serum) in suspending medium
- fast thaw from −196 °C to room temperature as rapidly as possible

Further consideration of the physico-chemical aspects of cell freezing is given in Ashwood-Smith & Farrant (1980).

4.2.1 *Equipment and materials*

The following is an outline of the practical aspects involved in freezing cell lines:

Sterile glass or plastic ampoules with a 1.5–2.0 ml capacity
Dispensing syringe of the automatic type

Sterile pipettes
Sterile dispensing vessel
Sealing torch for glass ampoules
Racks to hold the ampoules, preferably drilled aluminium
Programmable freezer or a two-stage freezer
Liquid nitrogen storage vessel
Improved Neubauer counting chambers
Trypan blue solution (0.4% w/v) in phosphate buffered saline
Cryoprotectant, such as growth medium with 20% serum and 7–10% cryoprotectant (DMSO or glycerol of the highest grade available)
Protective goggles, face mask and gloves

It should first be decided whether to store the cells in glass or plastic ampoules. Both types have certain advantages. Glass was used originally and is easy to clean and sterilise. Once properly sealed, glass ampoules can be kept indefinitely in liquid nitrogen. However, dangerous situations can occur if ampoules are improperly sealed; liquid nitrogen may leak in, so that on removal from the storage vessel the sudden expansion of liquid inside the ampoule can cause it to explode violently. Additionally, infectious microorganisms present in the nitrogen may enter the ampoule and lead to contamination of the culture.

The main problem, however, for anyone wishing to use glass ampoules, is the need for heat-sealing apparatus and ceramic ink to label the glass. Both require expensive equipment and a certain amount of technical expertise in use. The investment is worthwhile only if the majority of the ampoules are to be stored exclusively in the liquid phase in a refrigerator. If this system is used, pre-cleaned (acid-washed) ampoules should always be used to ensure that no toxic residues remain in the glass.

It is worth noting that the risk of ampoules exploding can be reduced by transferring their contents to the gaseous phase one or two days prior to when they are needed. This allows any liquid inside the ampoule time to 'leak-out' at a more controlled rate.

For most users plastic ampoules with screw-tops are more convenient. They are available pre-sterilised, are easy to handle and will label with permanent marking pens. A commercially available heat-shrinking sheath can be placed over the cap and the body of the ampoule to minimise entry of liquid into the ampoules. This enables cells to be stored at the more acceptable temperature obtained by immersion in liquid nitrogen, without the dangers arising with the use of glass ampoules. However, it may be decided to use the gaseous phase of nitrogen for storage, in which case great care will be necessary in the handling of the inventory storage system on entry to the refrigerator tank. Careless handling can lead to fluctuations in storage temperature and result in loss of viability of the frozen cells.

Another major decision is in the choice of freezing apparatus. The sophisticated, programmable type allows the greatest flexibility, as the cooling rate is fully controlled. However, they are expensive and only become a good investment in a laboratory that will be freezing cells regularly and in substantial quantities. A less expensive approach, especially applicable when small batches of ampoules are to be frozen, is the use of two-stage freezing. Primary freezing takes place in a specially designed Dewar flask which is filled with liquid nitrogen. Ampoules are suspended at a pre-determined depth in the nitrogen gas in the neck of the flask. By experimentation, an optimum cooling rate is found and once frozen the ampoules can be transferred to a storage vessel.

4.2.2 *Preparation and freezing of cells*

A fully optimised method for freezing cells will give the same percentage of viable cells on recovery as were present in the culture before cyropreservation. Outlined below are some basic steps to follow:

(i) Cell cultures must be healthy (by examination using an inverted microscope) and in log phase growth, and the growth medium not more than 2–3 days old at the time of freezing (i.e. not at nutrient exhaustion). A viability count is made using trypan blue, a vital stain which enters only dead cells. Preferably, viability should be higher than 95%, and cultures of 80% viability or below rejected.

Cultures should previously have been checked for microbial contamination, especially mycoplasmas (see section 4.2.3). In order to detect contamination easily cultures are grown without antibiotics.

(ii) When cultures are ready for freezing the cells are collected in the same manner as for routine subculture. Surface attaching cells are removed with the appropriate enzyme, e.g. trypsin or trypsin-EDTA. Enzymes must be neutralised by the addition of medium containing serum to a volume at least equal to that of the enzyme used.

Cells in suspension (e.g. hybridomas) are distributed into 50 ml or

250 ml sterile conical tubes for centrifugation. Cells should be centrifuged with the minimum force necessary to pellet them, thus reducing stress on the membranes which may be exacerbated during freezing.

A freezing mixture is freshly prepared from growth medium with 20% serum, to which 7–10% (v/v) DMSO has been added, or from whole serum, i.e. NBCS, with 7–10% cryoprotectant (DMSO or glycerol). The mixture is stored on ice. DMSO is a powerful solvent which can penetrate the skin, therefore care should be taken to avoid contact. If DMSO is used it is important to obtain the most recent stocks from a supplier and also not to use material that has been left standing on the shelf in the laboratory; this inevitably leads to toxicity problems due to oxidation. Ideally DMSO should be dispensed and stored at −20 °C prior to use.

(iii) The supernatant is decanted and the cells resuspended in the freezing mixture to a final concentration of $4–10 \times 10^6$ cells/ml^{-1}. The cells are then dispensed into ampoules. When using small volumes (1 ml per ampoule), it is simplest to use plastic pipettes. For larger volumes, an automatic syringe may be necessary. The vessel supplying the syringe should contain a teflon coated magnetic bar, so that it can be placed on a magnetic stirrer to keep the cells from settling. It is advisable to sterilise the syringe, tubing and vessel as one entity and not to try to assemble them separately afterwards, as this may lead to a greater risk of contamination.

If large volumes of freezing mixture containing growth medium with bicarbonate are to be dispensed a problem can occur with increased alkalinity. It may be necessary to gas the mixture with 5–10% CO_2/air as prolonged exposure to pH values higher than 7.7–7.8 may weaken or kill the cells. Normally the use of zwitterionic buffers (e.g. Hepes) is avoided because of toxicity problems.

Dispensing may be carried out in a laminar flow cabinet for normal primary cell lines (but not monkey kidney cells, due to the possible risk of virus contamination), but for transformed lines a class 2 containment cabinet is advisable. For lines carrying known or suspected pathogens it may be necessary to use category 3 facilities. This subject is further developed in Chapter 3.

(iv) If glass ampoules are used they must be sealed using an acetylene/oxygen or propane/oxygen torch. Unless an automated dispensing/sealing machine is used, manual sealing will require practice to achieve the smooth round tip necessary for a correctly sealed ampoule. After sealing, the ampoules are checked for leaks by immersing in a solution of 0.05% methylene blue at 4 °C and holding for 30–40 min. If any ampoules contain blue dye they are discarded and the remainder are frozen. This holding period allows some equilibration to occur between the cells and the freezing medium and enhances survival.

(v) For optimum survival during freezing an initial cooling rate of between 1–5 °C min^{-1} is suitable for most cell types. When a programmable freezer is available, a trial run at cool rates of between 2 and 3 °C min^{-1} is carried out for each cell line. At the end of the test run the contents from one ampoule are immediately resuscitated and the cells are grown for several days to assess their recovery capacity.

It is not necessary to maintain the same slow rate of cooling throughout the run; once the critical stages of cooling are past and a temperature of −60 °C has been reached, the rate can be increased to 10 °C min^{-1} to the final storage temperature of −196 °C.

The two-stage freezing system has an advantage over the programmable type in that ampoules can be placed at various levels in the Dewar flask, so providing different cooling rates during the same freeze run. This enables the optimum cooling rate to be established during the first testing period. There are also advantages of cost and minimum maintenance with this type of freezing equipment.

As most ampoules will probably remain in storage for long periods it is essential to maintain a good inventory of the location of each ampoule in the freezer. A record book should be kept with details of each freeze run and comments on resuscitated cultures. Many storage vessels have trays with symmetrical dividers so it is a simple matter to make a storage plan for each tray. When ampoules are withdrawn, the spaces left should be marked on the plan with a date. Larger laboratories often find it more convenient to hold inventory information on a microcomputer.

A typical large capacity liquid nitrogen refrigerator is shown in Fig. 4.1.

It must be stressed that when handling frozen ampoules, a full face mask and protective gloves must be used; ampoules should always be considered potentially explosive.

4.2.3 *Thawing and recovery of cells*

Thawing of cells is relatively straightforward providing certain basic procedures are followed.

(i) It is normal procedure to thaw ampoules in a 37 °C water bath. As

previously stated, all ampoules should be handled with extreme caution and a lid should be placed over the water bath for protection. Plastic ampoules should be placed in a rack and lowered into the water bath to a level just below the screw-cap. This reduces the risk of contamination from the water bath. If the ampoules contain potentially hazardous material, chloramine-T to a final concentration of 1–2% (w/v) can be added to the water as a precaution.
(ii) After thawing, the ampoules are transferred to a sterile cabinet. Glass ampoules are soaked in 70% ethanol and allowed to dry. Prescored ampoules can be snapped open with an alcohol-soaked tissue. Unscored ampoules must be scored at their waist with a diamond

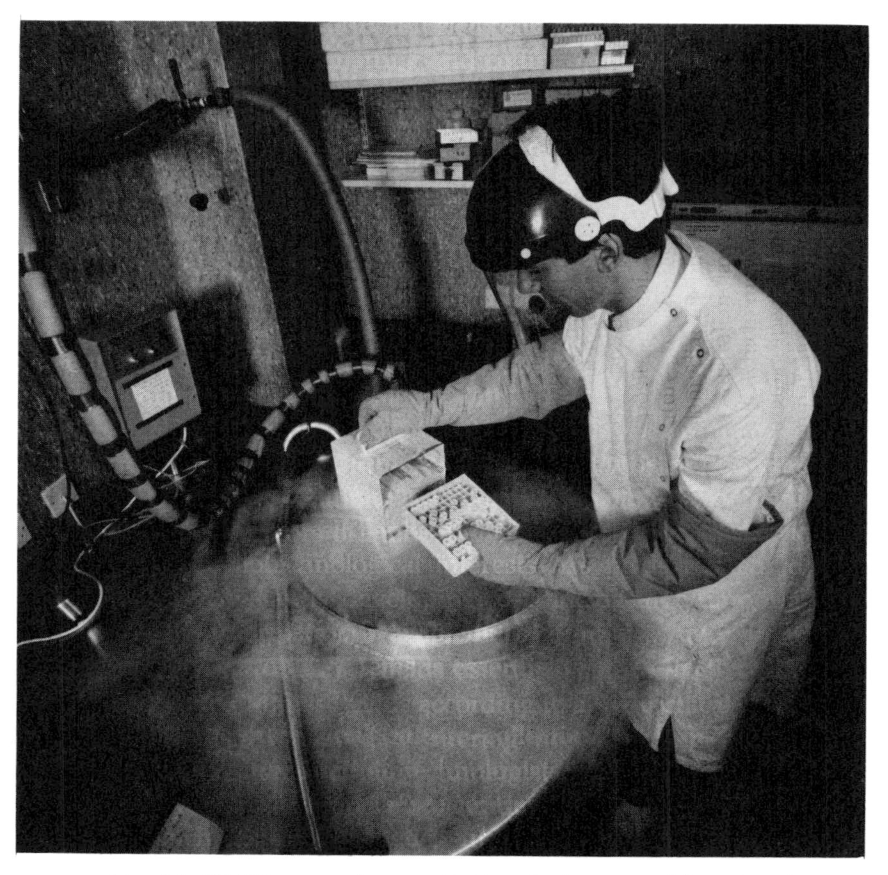

Fig. 4.1. 22 000 ampoule capacity liquid nitrogen refrigerator at ECACC.

cutter before soaking. They are then placed in a wooden or hard rubber block in which holes have been drilled to hold the ampoules. The top can be snapped off as for prescored ampoules, or alternatively safety ampoule breakers can be used. Plastic ampoules should be gripped with an alcohol-soaked tissue during opening to capture any aerosol released under pressure.

(iii) The contents are transferred with a sterile pipette to either a flask or centrifuge tube containing the appropriate growth medium at room temperature. As the speed at which the cryoprotectant is diluted out of the cells may be critical to their survival, the cells should be added to the growth medium in a dropwise manner. In certain cases (with particularly cryoprotectant-sensitive cells) it may be necessary to remove the cryoprotectant from the outset by centrifugation. The minimum centrifugal force needed to pellet the cells should be used. When a culture is grown directly from an ampoule, the medium is changed after the first 24–36 h to remove remaining DMSO.

(iv) The cell density at the time of resuscitation is critical. It is important to encourage the cells to condition the medium as rapidly as possible, and as the lag phase prior to steady growth will vary from cell type to cell type, a high starting density is desirable. In the case of most suspension cells densities of 3 and 6 $\times$ 10^5 cells/ml^{-1} are recommended; for surface attaching lines densities of 2–5 $\times$ 10^4 cells/ml^{-1} are appropriate. The viability of a culture can be tested by mixing the equivalent of 20–30 μl of the contents of an ampoule with a 5–10-fold dilution of trypan blue and counting the proportion of unstained cells under a microscope in an improved Neubauer counting chamber.

(v) It is important to examine cultures on an inverted microscope each day and subculture when confluent. During the first few days the cells are maintained at around the mid-log density. Adherent cells can be monitored with phase contrast microscopy – unhealthy cells will often have a more granular appearance. Suspension cells can be counted more readily and viability studies performed as necessary.

While the above methods cover the general points to be considered when freezing cell lines, it should be realised that many new types of cell have been cultured over the past decade, some of which require fastidious growth conditions for freezing, often including the need for growth factors such as Epidermal Growth Factor (EGF) or interleukins to be added to the medium. It may be conjectured that the association of the factor with its surface receptor may help to stabilise the membrane during freezing, or simply protect the integrity of the receptor for available growth factor after resuscitation.

Hybridomas often cause problems on resuscitation because of instability over the first few days, resulting in variable levels of dead cells. When the cell debris level becomes too high, it may induce further cell death because of the accumulation of toxic factors. This is a major problem when only a few million cells have been frozen. Starting the cultures on feeder layers of macrophages in high serum medium (20%) will not only provide rapidly conditioned medium in which the viable cells will be stimulated to divide, but cell debris will be endocytosed by the macrophages. Normally Balb/C peritoneal macrophages taken from one or two mice will be sufficient for several standard 24 well-plates. A concentration of 2×10^4 cells per well is recommended.

A general problem associated with the cryopreservation of cell lines is the pH stability of the medium during freezing and immediately after resuscitation. As previously mentioned it may be necessary when using bicarbonate buffered medium to make adjustments with carbon dioxide. However, during freezing there is a localised rise in ionic concentration of the extracellular solution as ice nuclei form. This has two effects. There is an increase in pH due to elevated levels of bicarbonate ions and also the CO_2 is forced out of solution. When the cells are resuscitated this effect will be exaggerated further because, as the temperature rises, there will be an increasing tendency for the gas to dissociate from the solution. At a time when membranes are at their most fragile, prolonged exposure to an alkaline pH may cause extensive damage.

One way to overcome the problem is to use whole serum with the cryoprotectant. Apart from offering a greater pH stability, the main components (globulin and albumin) have a limited buffering capacity and may provide a certain amount of physical protection. An additional advantage is the capacity of the serum to neutralise any traces of remaining trypsin. If newborn bovine serum is used rather than foetal bovine serum it may not have to be an expensive option.

The potential pathogenicity of cell cultures should always be considered. The problem of microbial contaminants, such as mycoplasma, has already been discussed and is covered in a later section (see p. 81). However, because the origin of many cell cultures is from sources that can include viruses, precautions in their handling must be taken as outlined in Chapter 3.

As cross-contamination from one cell line to another is a danger it is important to work with each cell line in isolation and keep all reagents separate. The increasingly rapid development of improved screening techniques undoubtedly reveals the presence of more contaminants in many more cell lines.

4.2.4 *Maintenance of hybridomas*

Among the cell lines of interest to the biotechnologist, hybridomas present particular problems. Detailed below are some of the factors which should be borne in mind regarding their long-term maintenance and culture.

Hybridomas are the result of the fusion of two cells. These cells may be of different types or even of different species. When fusion occurs between an antibody secreting cell from an immunised animal and a cell from a closely related malignant cell line then the resulting clone expresses the characteristics of both parents, becoming immortalised and secreting a single, unique antibody – a monoclonal antibody. Such antibodies have enormous commercial potential for diagnosis, immunotherapy and biotechnological processes.

Once a hybridoma is established, then bulk culture can be carried out, leading to large-scale production of the antibody (see p. 76). The routine maintenance of hybridomas is difficult. The process of hybridisation itself involves producing cells containing an extra complement of genetic material as an additional set of chromosomes. Such cells are inherently unstable and frequently lose complete chromosomes, or fragments of them. This loss may prevent the cell from producing the antibody which was of interest in the first place. In a mixed culture these non-producer cells quickly outgrow the remaining producer cells. It is, therefore, necessary to store, by cryopreservation, adequate stocks of cells at an early stage in their isolation and it is also important to test the cell supernatant at frequent intervals in order to ensure that antibody production remains constant.

The scientist is generally more concerned with the problem of selecting clones secreting suitable monoclonal antibodies rather than with their growth characteristics. Once a suitable hybridoma has been selected, however, efforts should be made to ensure that only a single clone is being grown. This is usually achieved by diluting the cells to a stage where only one cell grows in a single well of a multi-well plate. This 'cloning' procedure should be repeated several times to guarantee clonality. Fortunately, as an added benefit, this technique will also ensure that the final clone has strong growth characteristics. During subsequent culture it may be found necessary to reclone the hybridoma in order either to enhance or encourage special growth characteristics or reselect antibody secreting cells.

As mentioned previously, mycoplasma contamination is often a problem in the long-term maintenance of hybridomas and can result in poor growth and may inhibit antibody production. Although methods are available for the treatment of mycoplasma infection it is safer to return to an original, mycoplasma-free cryopreserved stock.

It should be remembered that each hybridoma is unique and as such may have special cultural requirements. It is unwise to consider them as a homologous group.

Analysing the specificity of monoclonal antibodies

The characterisation of monoclonal antibody specificity is an essential part of the production process. A bacterial cell, for example, may express thousands of antigenic molecules on its surface, each with multiple sites for antibody binding. It is not sufficient simply to establish that the monoclonal antibody recognises a bacterium. The type of antigen, its stability, cross-reactivity and the strength of binding (avidity), for example, will all need to be established before the antibody can be considered as a diagnostic reagent.

Many techniques are available to assist in identifying the antigen specific for the antibody. The enzyme-linked immunoassay system (ELISA), developed for hybridoma selection, may be used to identify the immunoglobulin class of the antibody and preliminary identification can be obtained if purified antigen is available. Alternative techniques include immunofluorescence and immunoperoxidase staining which can locate particular antigens in sections of tissue or in whole cells at the light microscope level, allowing correlation with populations or sub-populations of cells. Moreover, recently developed gold-labelled immunodetection systems may be used to observe monoclonal antibodies bound to antigens at the electron microscope level.

There are several specialised techniques available to analyse antigens at the molecular level. Mixed protein antigens may be separated into their component bands by polyacrylamide gel electrophoresis. Once separated, protein bands can then be transferred electrophoretically onto nitrocellulose paper, incubated with the monoclonal antibody and the antibody–antigen interactions observed. This technique is called Western blotting. Alternatively, mixed antigens in a soluble form can be radiolabelled with iodine (^{125}I). Following incubation with the antibody, the resulting complex can be captured onto Sepharose beads. The Sepharose beads are then washed free of unbound radiolabelled material and the bound ^{125}I-antigens separated by electrophoresis and detected by autoradiography. This procedure is called radioimmunoprecipitation.

All of these techniques are subject to the inconsistency of the monoclonal antibodies which are rarely positive in all assays. It is therefore essential to use a panel of techniques in characterisation and never to expect the monoclonal antibodies to have exactly the same properties as the antiserum.

Despite the original presumption that monoclonal antibodies would replace antisera and solve many diagnostic and biotechnological problems, they tend to present more questions than they answer.

4.3 Scaling-up of cell cultures

4.3.1 *Introduction*

Bulk production of animal cells has traditionally been required for vaccines, and more recently for interferon and monoclonal antibodies. Animal cells are far less productive than bacterial cells, and need more stringent culture techniques and more complex and expensive media. The fact that animal cells have an increasingly important role to play in biotechnology is a reflection of the fact that many biomolecules are not completely expressed in recombinant bacteria – not glycosylated, for example – or there are extraction and toxicity problems. In addition, the use of viral vectors such as Herpes simplex, vaccinia and baculoviruses (in insect cells) necessitates the use of animal cell cultures. To overcome the low productivity problem of animal cells, intensive studies are taking place to increase product yields by genetic and phenotypic amplification, by increased knowledge of cell physiology and the development of biosensors to control the environment and develop more efficient cell reactor systems. The problem of low productivity in the development of therapeutic agents is exemplified by the data on the volume of culture needed to produce clinical doses of some animal cell products, listed in Table 4.1 (Katinger & Bliem, 1983). The range of products, and potential products, for animal cell systems manufactured by recombinant technology is given in Table 4.2.

The normal growth characteristics of animal cells have led to the development of two completely separate systems for large scale culture. On the one hand, some cells can be grown as single discrete cells in agitated suspension culture in systems analogous to those used for bacteria; on the other hand, some cells are anchorage-dependent and will grow only when attached to a solid substrate. Thus scaling-up to bulk culture of animal cells has to follow two separate developmental paths to cater for both cell types. This separation into two groups is not distinct as many cells will grow in suspension, but with a much lower productivity of cell mass or product than when grown on a solid substrate.

4.3.2 *Principles of scale-up*

Transfer of cells from *in vivo* to *in vitro* conditions is a harsh treatment and can result in completely undifferentiated but fast growing cells. In small-scale culture extra attention to detail can be given to ensure that the cell is an optimum environment and thus behaves in a typical manner. During scale-up this extra attention cannot be provided so easily, and fastidious cells can be grown to bulk amounts only with difficulty. In addition, the threat of microbial contamination

Table 4.1. *Culture requirements to produce therapeutic doses of animal cell products*

Product	Cell requirement per dose	Culture volume (l)
Polio	2×10^4	0.0001
Rabies	4×10^6	0.005
HSV	2×10^7	0.03
FMDV	2×10^7	0.01
IFN (anti-viral)	10^5/day	0.1
(anti-tumour)	5×10^5/day	
t-PA	$>10^{10}$	1–10
MCAb	10^{12}	100
Urokinase	10^{12}	500

Data based on Katinger & Bleim (1983).

Table 4.2. *Recombinant products manufactured from animal cells*

Viruses	HBsAg, HSV 1 & 2, Influenza, CMV, EBV*, Rabies*, HIV*, FMDV*, Lassa fever*
Blood products	Factor VIII, Factor IX, Protein C, Immunoglobulins
Hormones	hGH, hCG, Insulin, Erythropoietin, Relaxin, LH
Others	t-PA, β-IFN, IL-2

* Expressed via Vaccinia virus.

increases with the extra manipulations (Arathoon & Birch, 1986) and use of more sophisticated and less easily sterilised equipment.

The principle of scale-up is gradually to change from multiple cultures to a unit process (Griffiths, 1986). This entails the design of suitable reactors and the development of systems to monitor and optimally control the environment. One of the principal limiting factors to scale-up is oxygen availability (Spier & Griffiths, 1984). Small cultures have a large surface area to volume ratio and this off-sets the low oxygen transfer rate (171 $\mu g\ cm^{-2}\ h^{-1}$) through the medium interface. Scaling-up results in a decreased surface area to volume ratio so that other means of maintaining adequate levels of oxygen are needed (Spier & Griffiths, 1984). The most efficient means of mass transfer of oxygen are high stirring rates and sparging (bubbling). However, both these methods are damaging to most animal cells which are very fragile compared with bacterial cells. Stirring at 300 rpm and sparging at 5 ml $min^{-1}\ l^{-1}$ are the usual maximum rates employed (Griffiths, 1986). Other factors that have to be controlled are pH, build-up of toxic metabolites such as ammonia and lactate, and nutrient exhaustion. Scale-up can be effected either by increasing the volume or by increasing the cell density per unit volume.

4.3.3 *Scale-up of suspension cells*

Spinner flasks, so called because they are stirred by a bar magnet driven by an external magnetic stirrer, may be used for volumes from 20 ml to 20 l (Griffiths, 1986). For most purposes, however, it is best to transfer to small-scale fermentation systems when volumes of 2–5 l are exceeded. This allows better control of stirring speed and mixing (by using more efficient impellers than bar magnets), pH and oxygenation. Stirred fermenters for animal cells are available up to 8000 l. An alternative system for fragile cells (such as many hybridomas) is the airlift reactor in which mixing is brought about by air bubbles rising up through a central draft tube (Birch *et al.*, 1985). These reactors are available in sizes from 500 ml to 1000 l. They are low process intensity systems with cell concentrations up to 3×10^6 ml^{-1} (Fig. 4.2).

Scale-up in cell density can be achieved in a variety of ways, including encapsulation (Rupp, 1985), membrane reactors (Klement, Scheirer & Katinger, 1987) and hollow fibre cartridges (Tharakan & Chau, 1986) (Fig. 4.1). Yields of over 10^8 cells ml^{-1} can be achieved in these units by means of medium perfusion from an environmentally controlled

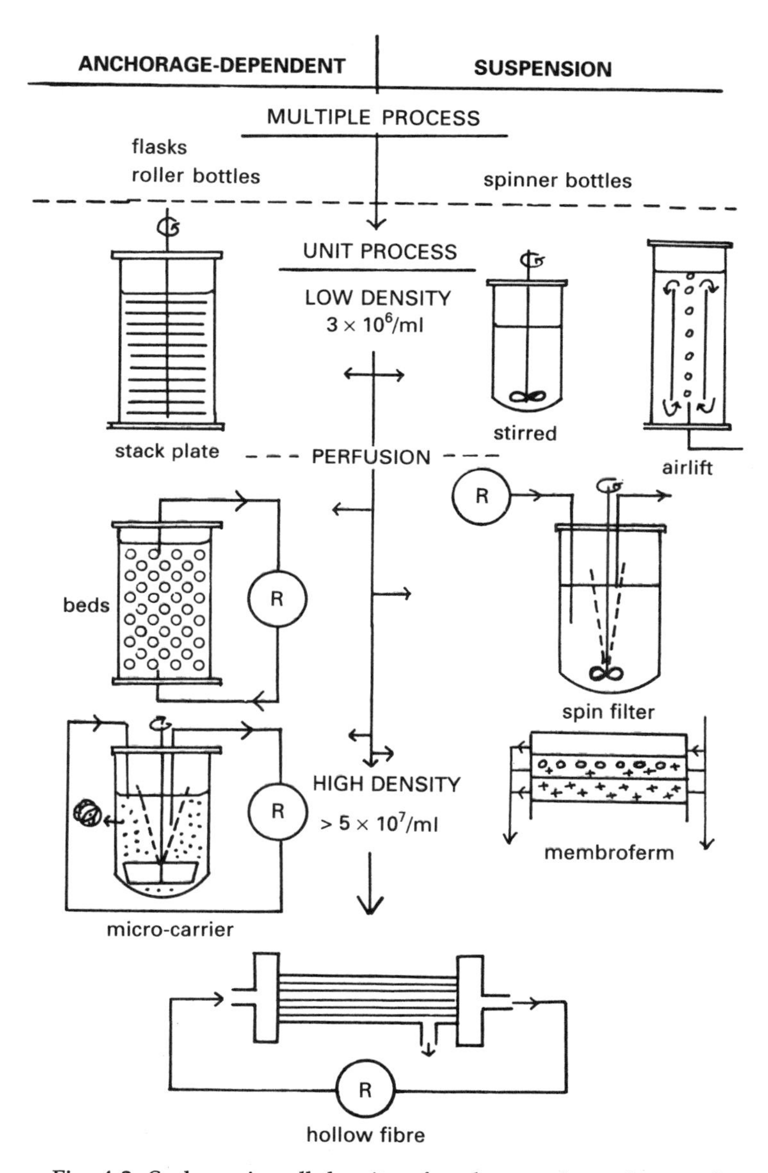

Fig. 4.2. Scale-up in cell density of anchorage-dependent and suspension cells. Diagrammatic representation of low intensity fermenters (stirred and airlift), the first stages of scale-up using spin filter techniques, and the high density systems using flat or hollow fibre membranes.

reservoir. As 1 ml of medium supports only about 1×10^6 cells the reservoir size has to be in the order of 100 times the culture volume; thus in reality these systems still need large volume culture equipment. An advantage is that the cell product is compartmentalised in a small volume by a semi-permeable membrane so that high yields can be harvested, thus facilitating downstream processing.

4.3.4 *Scale-up of anchorage-dependent cells*

Originally, scale-up of animal cells was made by using a multiplicity of small units such as flasks and roller cultures. To reduce the number of vessels involved various modifications to these units have been made to increase the surface area in proportion to the total volume (Griffiths, 1983, 1986). This in turn has led to the use of bulk culture systems such as stack plates, multitrays and high surface area matrices (Spier, 1985). Another approach has been to use immobilised beds, and glass spheres have been widely used for this purpose (Whiteside & Spier, 1981). A third approach has been the microcarrier system in which cells are grown on small spheres which are kept in stirred suspension in conventional fermentation equipment with all the process control advantages of a suspension system (Reuveny, 1983).

All the developments have been in response to the need dramatically to increase the available surface area in relation to the culture volume so that a unit, rather than a multiple, process could be used. The need to increase the cell density per unit volume of medium is a more recent development. This is to compensate for the approximately 100-fold lower productivity of animal cells compared with microorganisms. The aim is to mimic *in vivo* cell densities ($>10^8$ ml^{-1}) and produce a high concentration of product to reduce downstream processing problems. This has not been achieved to the same degree as for suspension cells but perfusion of microcarriers by spin-filter techniques (Griffiths *et al.*, 1984; Feder & Tolbert, 1985) allows densities in the region of $1–5 \times 10^7$ cells ml^{-1} to be achieved.

4.4 References and further reading

Storage and maintenance

Ashwood-Smith, M. J. & Farrant, J. (ed.) (1980). *Low Temperature Preservation in Medicine and Biology*. Tunbridge Wells, Kent: Pitman Medical Ltd.

Freshney, R. I. (1983). *Culture of Animal Cells: A Manual of Basic Technique*. New York: Alan Liss.
Freshney, R. I. (ed.) (1986). *Animal Cell Culture: A Practical Approach*. Oxford: IRL Press.
Goding, J. W. (1983). *Monoclonal Antibodies: Principles and Practice*. London: Academic Press.
Langone, J. & Van Vunakis, H. (1986). *Methods in Enzymology, 121, Immunochemical Techniques, Part I, Hybridoma Technology and Monoclonal Antibodies*. Orlando: Academic Press.
Mishell, B. B. & Shiigi, S. M. (1980). *Selected Methods in Cellular Immunology*. San Francisco: W. H. Freeman & Co.
Weir, D. M. (1986). *Handbook of Experimental Immunology, 4, Applications of Immunological Methods in Biomedical Sciences* (4th edn). Oxford: Blackwell Scientific Publishers.

Scaling up

Arathoon, W. R. & Birch, J. R. (1986). Large-scale cell culture in Biotechnology. Science **232**, 1390.
Birch, J. R., Thompson, P. W., Lambert, K. & Boraston, R. (1985). The large-scale cultivation of hybridoma cells producing monoclonal antibodies. In *Large-Scale Mammalian Cell Culture*, ed. J. Feder and W. R. Tolbert, pp. 1–18. Academic Press.
Feder, J. & Tolbert, W. R. (1985). Mass culture of mammalian cells in perfusion systems. *Int. Biotech. Lab.*, June, 40.
Griffiths, J. B. (1983). Animal cell cultures in medicine. *Lab. Practice* **32** (12), 11.
Griffiths, J. B. (1986). Scaling-up of animal cell cultures. In *Animal Cell Culture: A Practical Approach*, ed. R. I. Freshney, Chapter 3, 33. Oxford: IRL Press.
Griffiths, J. B., Atkinson, R., Electricwala, A., Latter, T., Ling, R., McEntee, I., Riley, P. M., Sutton, P. M. (1984). Production of a fibrinolytic enzyme from cultures of guinea pig keratocytes grown on microcarriers. *Develop. Biol. Standard.* **55**, 31.
Katinger, H. W. D. & Bliem, R. (1983). Production of enzymes and hormones to mammalian cell culture. *Adv. Biotech. Proc.* **2**, 61.
Klement, G., Scheirer, W. & Katinger, H. W. D. (1987). Construction of a large scale membrane reactor system with different compartments for cells, medium and product. *Develop. Biol. Standard.* **66**, 221.
Reuveny, S. (1983). Microcarriers for culturing mammalian cells and their applications. *Adv. Biotech. Proc.* **2**, 2.
Rupp, R. G. (1985). Use of cellular microencapsulation in large-scale production of monoclonal antibodies. In *Large-Scale Mammalian Cell Culture*, ed. J. Feder and W. R. Tolbert, pp. 19–38. Academic Press.
Spier, R. E. (1985). In *Animal Cell Biotechnology*, Vol. 1, ed. R. E. Spier and J. B. Griffiths, p. 243.
Spier, R. E. & Griffiths, J. B. (1984). An examination of the data and concepts germane to the oxygenation of cultured animal cells. *Develop. Biol. Standard.* **55**, 81.

Tharakan, J. P. & Chau, P. C. (1986). A radial flow hollow fiber bioreactor for the large-scale culture of mammalian cells. *Biotechnol. Bioeng.* **28**, 329.

Whiteside, J. P. & Spier, R. E. (1981). The scale-up from 0.1 to 100 litres of a unit-process system based on 3 mm diameter glass spheres for the production of 4 strains of FMD from BHK monolayer cells. *Biotech. Bioeng.* **23**, 551.

5

Quality control

A. DOYLE, C. MORRIS and J. M. MOWLES

5.1 Introduction

An essential part of the routine procedures associated with tissue culture laboratories concerns quality control. Details of techniques are given which relate to the routine monitoring of cell lines for microbiological contamination and also species and other identification. In routine handling of cell lines these are the techniques most commonly used. Other tests are required before cell lines can be accepted for the production of proteins for therapeutic use, for example, and there are guidelines laid down by the US Food and Drug Administration and the Commission of the European Communities (CEC) on the nature and extent of the recommended test procedures; although these are beyond the scope of this volume, further details may be obtained from Esber, 1987 and CEC, 1988.

5.2 Testing for mycoplasma

Mycoplasma contamination of cultured cells was first identified in 1956 (Robinson, Wichelhausen & Roizman, 1956). Many reports have followed this original observation illustrating the frequent occurrence of this type of contamination in cell cultures. Approximately 15% of the cultures received by one cell culture bank (ECACC: see Chapter 1) are mycoplasma infected. The majority of cell culture mycoplasma infections are caused by five species. These and their natural hosts are *M. arginini* (bovine), *M. fermentans* (human), *M. hyorhinis* (porcine), *M. orale* (human) and *Acholeplasma laidlawii* (bovine). The genus *Acholeplasma* differs from other members of the order Mycoplasmatales by its non-dependence on sterol for growth.

Because these organisms often do not exert any obvious effect upon the well-being of the infected cell culture, many investigators regard them as merely a nuisance and have ignored their possible role in experimental results. However, amongst other effects, mycoplasmas have been found to interfere with the growth rate of cells (Barile & Levinthal, 1968; Simberkoff, Thorbecke & Thomas, 1969; Callewaert *et al.*, 1975; McGarrity, Phillips & Vaidya, 1980), induce morphological alterations, including cytopathology (Kraemer *et al.*, 1963; Butler & Leach, 1964), induce chromosome aberrations (Paton, Jacobs & Perkins, 1965; Aula & Nichols, 1967; Fogh & Fogh, 1967), affect amino acid (Stanbridge, Hayflick & Perkins, 1971) and nucleic acid metabolism (Levine *et al.*, 1968), and cause membrane alterations (Wise, Cassell & Acton, 1978) and cell transformation (MacPherson & Russell, 1966; Kotani, Phillips & McGarrity, 1986).

A wide variety of assay techniques to detect mycoplasma infection of cell cultures is available and has been reviewed elsewhere (Hessling, Miller & Levy, 1980; McGarrity, 1982; McGarrity, Kotani & Carson, 1986). Where possible at least two procedures should be used to minimise false-positives and false-negatives on testing. For routine application, DNA staining and microbiological culture are appropriate methods which most laboratories should be capable of performing.

Prior to testing, cultures should be grown for at least two passages in antibiotic-free medium as the presence of antibiotics may mask infection (McGarrity, Sarama & Vanaman, 1979). Cell cultures should not be tested from thawed ampoules until two passages are complete, as cryoprotectants may also mask infection. In addition cultures should not receive a medium renewal within 3 days of testing, to avoid dilution of low levels of contamination.

5.2.1 *DNA staining*

The fluorochrome dye, Hoechst 33258, binds specifically to DNA. Uninfected cell cultures observed under fluorescence microscopy are seen as fluorescing nuclei against a negative background, whereas cultures infected with mycoplasma contain both fluorescing nuclei and extranuclear mycoplasmal DNA. The main advantages of the staining method are the speed at which results are obtained and that non-cultivable mycoplasma strains (Hopps *et al.*, 1973) can be detected.

Method

The method was originally described by Chen (1977) with adaptations by Barile (1977) and is described here with other amendments.

Sterile coverslips are aseptically placed into 35 mm culture dishes, onto which is seeded an indicator cell line such as Vero (African Green Monkey Kidney) at a concentration of approximately 10^4 cells ml^{-1} in antibiotic-free media. After overnight incubation in 5% CO_2:95% air at 37 °C the cultures are examined microscopically to ensure attachment and viability of cells and absence of bacterial or fungal contamination. The test is carried out by adding test samples at 5×10^5 cells ml^{-1} to each of two prepared culture dishes. Two negative controls to which no test cells are added are also included. Where practicable positive mycoplasma cultures such as *M. hyorhinis, M. orale* or a known positive cell line should also be included in the test procedure. All cultures are reincubated for 8–12 h. A coverslip of one negative and, if included, one positive culture is taken for an initial test, while reincubating the remainder for at least a further 48 h. Prior to fixing the cells for staining, a further microscopic examination should be made for bacterial and fungal contamination.

Coverslip cultures are fixed with freshly prepared Carnoy's fixative (3:1 methanol:glacial acetic acid) and air-dried for 30 min or until dry. The fixed preparations are stained with Hoechst 33258 at a final concentration of 0.05 μg ml^{-1}, mounted and examined by UV epifluorescence at 100× magnification for the presence of mycoplasma.

Small fluorescent cocci or filaments in either of the test specimens are indicative of a mycoplasma infection (Figs. 5.1 and 5.2). The test is invalidated if mycoplasma either are detected in the negative control samples or are not detected in the positive control sample.

The use of an indicator cell line is considered desirable to promote better standardisation and allow appropriate positive and negative controls to be included in each assay.

5.2.2 *Microbiological culture*

Nearly all mycoplasma cell culture contaminants will grow readily on standardised agar and broth media (with the exception of certain strains of *M. hyorhinis*). The agar isolation technique remains an effective means for the isolation and subsequent identification of mycoplasmas from infected cells.

Method

For routine day-to-day use, commercial culture media are available and, although not as effective for isolation as those prepared from basic ingredients, they provide a satisfactory alternative for laboratories that do not wish to become specialist mycoplasma research departments. The type of isolation media available varies from ready prepared flasks containing both agar and broth such as Mycotrim (Hana Biologics),

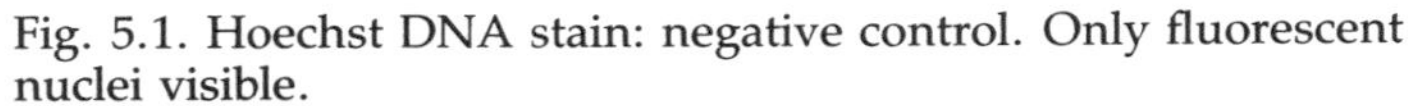

Fig. 5.1. Hoechst DNA stain: negative control. Only fluorescent nuclei visible.

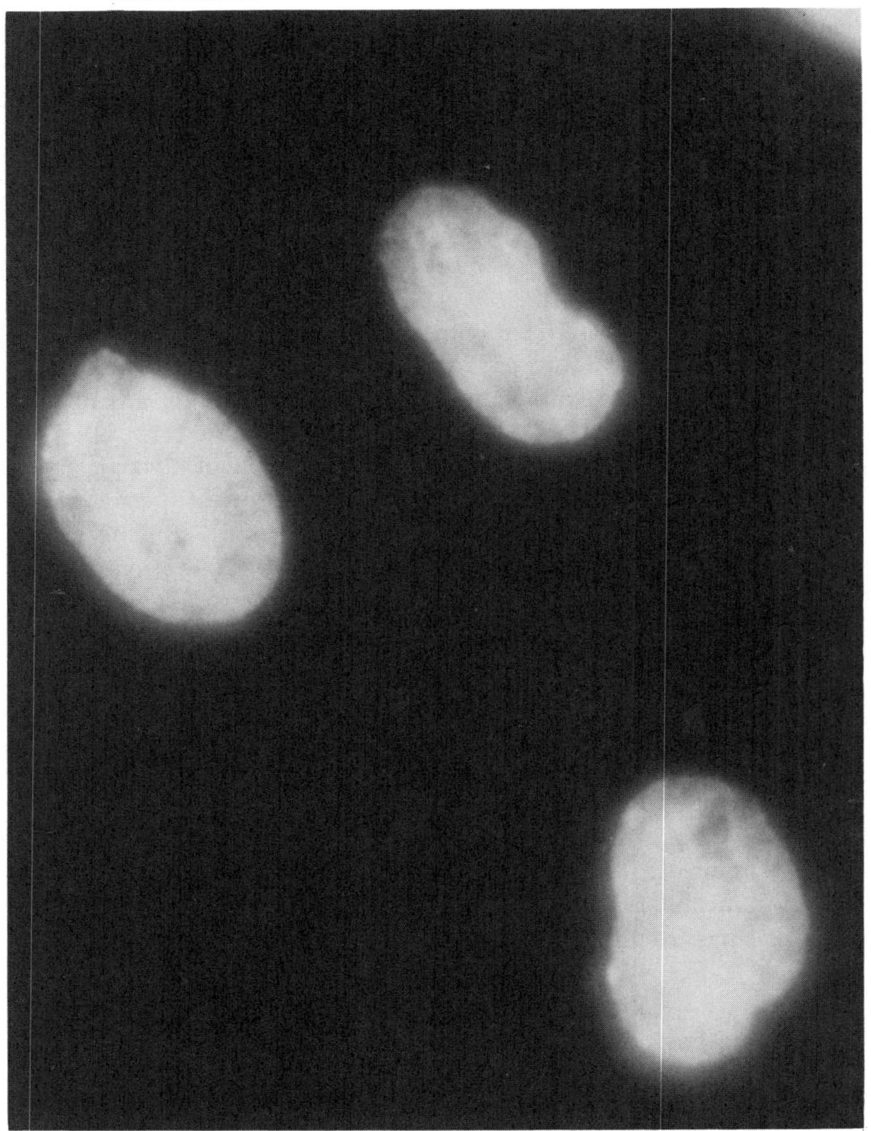

basal agar media to which supplements are added (Wellcome) or dehydrated media with supplements (Oxoid). The constituents for easily prepared media formulations are given below:

Fig. 5.2. Hoechst DNA stain. Fluorescent nuclei and mycoplasma visible.

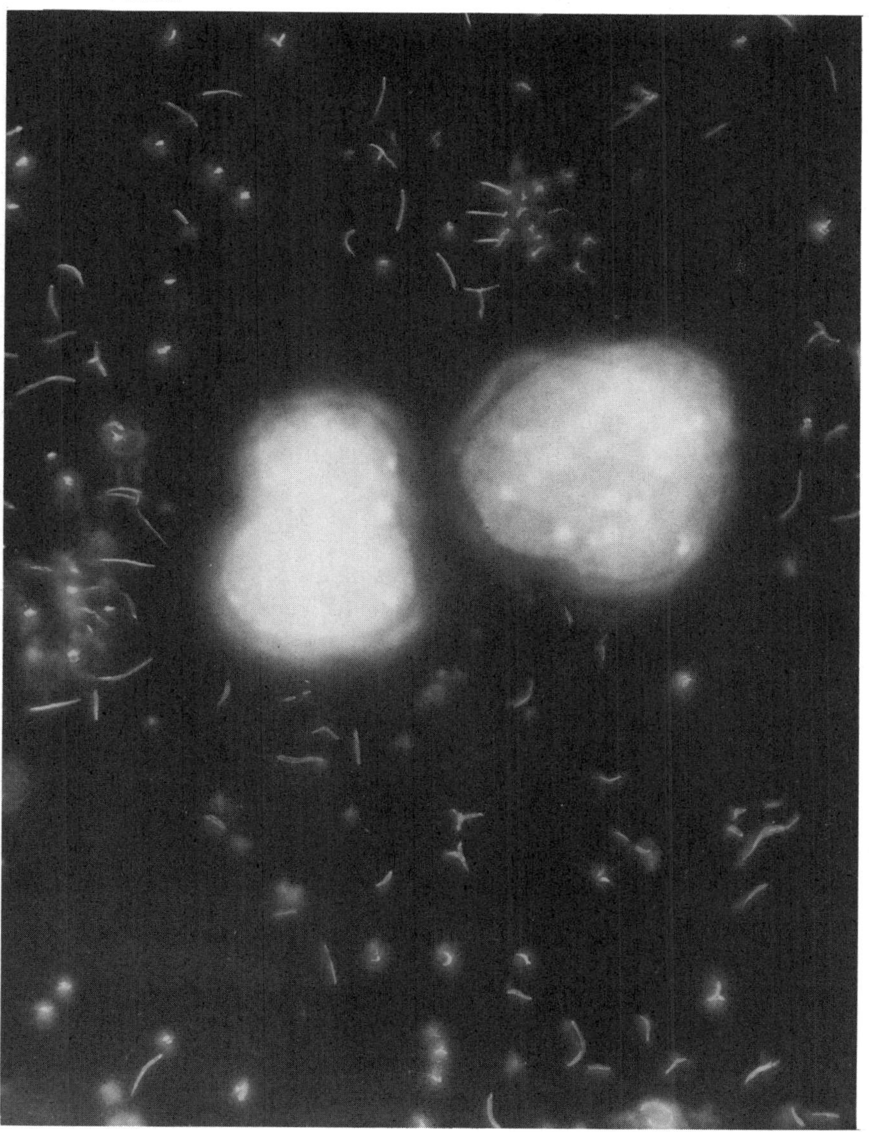

Agar medium:

Oxoid Mycoplasma agar base 80 ml
Pig serum (heat inactivated, 56 °C for 45 min) 10 ml
Oxoid yeast extract (0.7 g to 10 ml water) 10 ml
Autoclave agar, cool to 50 °C and mix with other constituents previously warmed to 50 °C.

Dispense 8 ml medium into each 5 cm diameter Petri dish. Complete medium should be used within 10 days. Store at 4 °C in sealed plastic bags.

Broth medium:

Oxoid Mycoplasma broth base 70 ml
Horse serum 20 ml
Oxoid yeast extract (0.7 g to 10 ml water) 10 ml
Phenol red, filter sterilised (0.2 μm) 0.002%
L-arginine, filter sterilised (0.2 μm) 0.2%
D-glucose, filter sterilised (0.2 μm) 1.0%
Adjust pH to 7.6

1.7 ml aliquots are dispensed in glass vials and stored at 4 °C. Complete medium may be stored for several weeks without deterioration.

Inoculation

Monolayer cell cultures are harvested using a scraper and resuspended to a concentration of approximately 5×10^5 cells ml^{-1} in the original medium. Proteolytic enzymes such as trypsin should not be used as they are detrimental to mycoplasmas (McGarrity *et al.*, 1979) and lead to false negative results.

Cells in suspension (e.g. hybridomas) are taken from culture at a density of approximately 5×10^5 cells ml^{-1}.

One agar plate is inoculated with 0.2 ml of the test sample and two mycoplasma broths are inoculated with 0.5 ml of cell suspension. A positive control, inoculated with a known mycoplasma cell contaminant, such as *M. hyorhinis* or *M. orale*, and a negative control, which is uninoculated, should also be included. Agar plates are incubated at 37 °C anaerobically (95% N_2/CO_2) (McGarrity & Coriell, 1973) and mycoplasma broths aerobically at 37 °C. Agar plates are examined for growth using an inverted microscope (100 × magnification) at 3, 7, 14, 21 and 28 days after inoculation. Mycoplasma broths are observed daily for turbidity and change in indicator colour which would denote a pH change following growth of contaminants.

At 7 and 14 days after incubation, 0.2 ml broth is transferred to agar plates. These plates are incubated anaerobically and examined on an inverted microscope at weekly intervals for two weeks. In addition, any broths which show turbidity or undergo a pH change are subcultured onto agar plates. It should be noted that some cell culture specimens can produce a pH shift in mycoplasma broth and that a change in pH is not necessarily diagnostic of mycoplasmas if an agar subculture proves negative.

It is necessary to be able to distinguish 'pseudocolonies' and cell aggregates from mycoplasma colonies on agar. 'Pseudocolonies' are caused by crystal formation and may even increase in size. They can be distinguished from genuine mycoplasma by using Dienes stain (Razin, 1983), which stains mycoplasma colonies blue but does not stain pseudocolonies. Cell aggregates, on the other hand, do not increase in size and so are more easily distinguished. By using a sterile bacteriological loop, cell aggregates may be dispersed leaving the agar surface free of aggregates. Mycoplasma colonies, however, due to the nature of their growth, will leave a central core which is embedded in the agar. Another good indication of genuine mycoplasma colonies is their typical 'fried-egg' appearance on agar; however, this is not always seen in primary isolates.

5.2.3 *Other methods of detection*

Two commercially available detection methods recently introduced are the MycoTect system (Bethesda Research Labs) and Gen-Probe (Gen-Probe), both of which are indirect methods of detection.

The MycoTect system requires co-cultivation of test cells with 6-methylpurine deoxyriboside (6-MPDR), which is non-toxic to mammalian cells. However, this is converted by mycoplasma phosphorylase to 6-methylpurine and 6-methylpurine riboside, both of which are toxic to animal cells and therefore destroy them. Results are available within 3–4 days. For a comparison of this method with Hoechst staining and culture, see McGarrity *et al.* (1986).

Gen-Probe is a DNA hybridisation assay which requires no cultural procedure and claims to give results within an hour. The protocol requires incubation of a ^{3}H-labelled DNA probe with either cell culture supernatant or the cells themselves. Hydroxyapatite beads are used to separate bound from unbound probe prior to scintillation counting. The DNA probe used is homologous to mycoplasmal r-DNA and therefore hybridises with different species of mycoplasma, but not with mammalian cellular or mitochondrial r-RNAs.

5.3 Testing for bacteria and fungi

Both cell cultures and cell culture reagents are at risk from contamination by an enormous range of organisms naturally occurring in the environment. Where sources of cell culture infection have been traced, these have included serum, trypsin, water baths and a hand lotion dispenser (McGarrity, Vanaman & Sarama, 1978). Organisms isolated from cell cultures include species of *Pseudomonas* and *Staphylococcus*, diphtheroids and *Escherichia coli* among the bacteria and *Penicillium*, *Aspergillus* and *Candida* among the fungi.

As mentioned before, prior to routine bacterial and fungal tests being performed cultures are grown in antibiotic-free medium for at least two passages. Antibiotics could easily mask infection and lead to a false-negative test result. In some circumstances this removal of 'blanket' antibiotic cover alone may prove sufficient to reveal infection.

Method

A wide range of commercial media is available for the growth of bacteria and fungi. The two types of isolation media suggested by the United States Pharmacopeia (USP, 1985) and by the European Pharmacopeia (EP, 1980) as being suitable for revealing the presence of viable forms of bacteria, fungi and yeasts are Fluid Thioglycollate Medium and Soybean-Casein Digest (or Tryptone Soya Broth). These dehydrated media are available from Oxoid and Difco. The technique outlined below is suitable for bacterial and fungal testing of cell cultures and cell culture reagents.

Broth is prepared and dispensed in 15 ml aliquots in glass universal bottles and stored at 4 °C. Once prepared, media may be kept for several months without deterioration.

Monolayer cell cultures are harvested using a scraper and resuspended to a concentration of approximately 5×10^5 cells ml^{-1} in the used culture medium. Proteolytic enzymes should not be used at this stage, as they could be detrimental to the isolation of bacteria and fungi.

Suspension cell cultures are sampled directly from the flask at a density of approximately 5×10^5 ml^{-1}. A test sample (1 ml) is inoculated into each of two Fluid Thioglycollate Medium broths and two Soybean-Casein Digest broths. Where possible, and as recommended by the USP and EP, positive controls should also be prepared with 1 ml inoculum each of a broth culture of *Bacillus subtilis* (ATCC 6633) and *Candida albicans* (ATCC 1031) inoculated into parallel samples of the same isolation media to ensure their ability to grow each type of organism. In addition, two of each broth type are left uninoculated as negative controls. Duplicate samples are incubated at 26 °C and 37 °C and are observed each working day over two weeks for signs of growth. Any broths which become turbid or display other changes may be subcultured to solid agar to assist in identification.

Eradication of microbial contamination

Wherever practicable cell cultures must be grown in antibiotic-free conditions as their constant use may lead to hidden infection by organisms which are resistant to the antibiotic used. The use of antibiotics should not take the place of good aseptic technique. Indeed some antibiotics can damage cell lines permanently by their effects on the chromosomes (Neftel *et al.*, 1986).

Where bacterial or fungal contamination is suspected (for example by a rapid change in pH or by a cloudiness in the medium) a short-term period of antibiotic treatment may prove effective in eradicating the infection. If facilities are available to identify the organism, at least to Gram stain status, a suitable single antibiotic may be used, such as penicillin against Gram-positive organisms or streptomycin against Gram-negative organisms. Where Gram staining is not possible a mixture of antibiotics may be necessary.

Fungal contamination is usually obvious either by the presence of mycelial growth or yeast forms in the medium and may be treated, if found early enough, with amphotericin B or nystatin.

Unlike bacterial or fungal contamination mycoplasma infection is not immediately evident in a cell culture by the usual microscopical methods. In some cases cells may fail to thrive; in other cases an infection may be present with no outward signs, emphasising the need for continual monitoring (at least once a month) of cell lines. Methods of eradication of mycoplasma from cell cultures have included passage in athymic, nude mice (Van Diggelen, Shin & Phillips, 1977), growth of cells in rabbit or guinea pig serum (Nair, 1985), and the use of nucleic acid analogues (Marcus *et al.*, 1980) or antibiotics (Schmidt & Erfle, 1984). For most laboratories the most convenient method is antibiotic treatment.

Various antibiotics have been investigated for their ability to eliminate mycoplasmas from cell lines, the most effective at present being tiamulin and minocycline (Schmidt & Erfle, 1984) and a new fluoroquinolone antibiotic, ciprofloxacin (Mowles, 1988). A combination of two antibiotics, used sequentially and marketed as BM Cycline (Boehringer Mannheim) have been prepared specifically to aid in the elimination of mycoplasmas. Recent reports (Mowles, 1988) indicate that ciprofloxacin and other 4-quinolone antibiotics are very effective in mycoplasma eradication.

Once antibiotic treatment has been initiated each passage should be performed at the highest dilution of cells at which growth occurs. After three or four passages the culture should be screened for mycoplasma. If mycoplasmas are still detectable after this period it is unlikely that the antibiotic used has proved successful and treatment with another antibiotic should be carried out. However, lack of evidence of mycoplasma does not necessarily indicate that the culture is free of infection, since the level of infection may be below the limit of detection. It is therefore suggested that antibiotic treatment is continued for at least five passages. After this period a further 10 passages without antibiotics should be performed to allow any residual infection to reach levels which are detectable. If after this period no mycoplasmas are detected, the line may be considered to be mycoplasma-free.

The following procedures are recommended to reduce the incidence of microbial infection:

(1) obtain cells from a reliable source
(2) carry out quality control tests on cells both on arrival and routinely during culture
(3) carry out quality control tests on media and reagents prior to use
(4) disinfect work-tops and wash hands between working with different lines
(5) never use the same media bottle for different lines
(6) use antibiotic-free media unless contamination is suspected
(7) establish good communications between technicians and supervisors. It is often the individual who routinely subcultures a line who first notices changes in a culture.

5.4 Identification

5.4.1 *Isoenzyme analysis*

As an integral part of its function it is important that a culture collection is able to verify the species of the various organisms it holds. In the case of animal cells, a karyotype for each species provides the necessary identification and also some background on the history of the cells. However, should a rapid identification be needed, the traditional procedures involved in obtaining a karyotype usually take too long to perform. A quicker molecular approach has now been developed using the unique isoenzyme profile provided by individual species.

Typically, an isoenzyme can be thought of as polymorphism within a specific protein structure. To be a true isoenzyme it must catalyse the same chemical reaction, but may exhibit small composite changes within its primary or tertiary structure. These changes will affect the protein's electrophoretic mobility and so provide a basis for identification.

Heterogeneity of the isoenzymes can occur at various points along the synthetic pathway. The main causes for alteration are multiple loci, multiple allelism and post-translational modification. One or all of these may be involved in enzyme heterogeneity. For further reading the excellent handbook of enzyme electrophoresis in human genetics by Harris & Hopkins (1976) is recommended. As illustrations of the extent to which cross-contamination within cell cultures can occur, the publications of Nelson-Rees, Daniels & Flandermeyer (1981) and Halton, Peterson & Hukku (1983) are suggested reading.

The need for a reliable and fast method of identification has in part arisen from an awareness that because different lines are handled in the same laboratory some interspecies contamination can occur, as illustrated by the presence of Hela markers in many cell lines. Also the transfer of cell lines from one laboratory to another carries the danger that the wrong cells can be transferred at some stage unless the line is carefully monitored. With long-term projects, often leading to commercial end-points, the need to verify a cell line's origin becomes imperative if precious resources are not to be wasted. Increasing interest in cell line identification has been shown by the US Food and Drug Administration and other regulatory bodies.

Originally, isoenzyme separations were performed on a starch-gel matrix. The resultant gel was sliced and the appropriate substrate applied in the presence of a formazan precipitating dye to mark the site of activity. This system required moderate expertise, especially in the cutting and staining of the gels because of their extreme fragility. Interpreting the complexity of bands also needed practice.

A few years ago a simplified system was developed by Innovative Chemistry Inc. (Marshfield, USA) using agarose as the gel support. By placing them on a flexible plastic support it is possible to produce gels of around 0.5 mm thickness. This affords several advantages. The running time is much reduced (to around 30 min), very little sample is needed and the gel can be dried after staining to provide a permanent record. Due to the speed with which the samples can be processed, it is possible to think of 'same day results' and the elimination of the prolonged wait involved in karyotyping.

Sufficient material can be extracted from one 25 cm^2 flask of cells to run several enzymes, as a sample of only 1 μl will give a result. The cell extract is stabilised in a buffer which allows the isoenzymes to be stored almost indefinitely. For comparative purposes it is necessary to have a standard and control as a source of reference. These can be supplied with the kit from mouse and human cell lines. Sample slots are preformed thus facilitating easy loading and the enzyme substrates are provided in a lyophilised form.

The choice of enzymes will to a certain extent reflect the need to identify the supposed species against known species. In many cases a good initial test enzyme is lactate dehydrogenase, as it has a tetrameric structure and can give up to five distinct bands, which in themselves may provide sufficient information to distinguish certain species. When identification involves primate species or rodents, nucleoside phosphorylase provides a clearer separation. In certain cases the difference between human lines derived from Caucasian and Negro origin can be seen with glucose-6-phosphate dehydrogenase. This can also be of help in distinguishing Hela contamination, because type B (found in 20% of Negroids) enzyme is present in this cell line. Another useful enzyme is peptidase B, as the differential migration rates between species is widely spread.

By using several enzymes a composite picture, analogous to finger printing, is built up of the species of origin. Should the cell line consist of a mixture from two species this can usually be seen, provided the minimum level of either is about 20% of the total cell number.

Experience has shown that this method of validation is invaluable for assessing cell lines. Cases of misidentification can readily be detected. However, this technique is of less help in detecting contamination by cells of the same species of origin.

5.4.2 *Cytogenetic analysis*

Cytogenetic techniques have been successfully used in a wide variety of areas over many years, most notably in establishing the normal karyotype of species and cell lines and then in identifying changes to them. In particular, they have been made to show changes associated with transformation and neoplasia, for identification of parental contribution of chromosomes in somatic cell hybrids, for gene mapping, verification of cross-contamination of cell lines, prenatal diagnosis in human medical genetics and in the quality control of cell lines used in the production of vaccines.

Techniques available

Conventional staining. Originally the technique was one of conventional or solid staining using either Giemsa or orcein to obtain an even staining of the chromosomes. This permitted karyotyping on the basis of chromosome morphology and position of the centromere. The technique has now become largely obsolete since the introduction of chromosome banding. It can, however, be very useful for studies on chromosome damage as scoring gaps and breaks can be difficult to detect in lightly stained chromosome bands. For example, this type of analysis is used as part of the quality control process of cell lines used in vaccine production and in the screening of chemicals when looking for clastogenic activity.

Giemsa banding. Although many species have distinctive karyotypes and can be readily identified by conventional staining with Giemsa or orcein it is essential to have some means of distinguishing between, for example, closely related species or tumour cell lines. This is provided mainly by G-banding which, using trypsin and Giemsa, gives rise to banding patterns on the chromosomes (G-bands) which are characteristic of each chromosome pair. This permits recognition by an experienced cytogeneticist of comparatively minor inversions, deletions or translocations. Other banding techniques are used in conjunction with G-banding, usually to clarify an area of concern in the overall pattern of results.

Several methods are available for G-banding, two of the most commonly used techniques being the pretreatment of slides with trypsin (Seabright, 1972) and the ASG technique (Sumner, Evans & Buckland, 1971), incubating the slides in a saline citrate solution at 65 °C as a pretreatment.

Quinacrine banding. Quinacrine banding (Q-banding) was one of the first banding techniques to be developed (Caspersson, Lomakka & Seck, 1971) and uses quinacrine mustard or quinacrine dihydrochloride to produce characteristic bright and dull fluorescent bands, which correspond, with few exceptions, to those of G-bands. Although it is a quick and simple technique for producing chromosome bands with the advantage that the slides can be used for a second or even third staining technique, G-banding is the preferred banding technique. The main reason for this is that the fluorescence does not last long and analysis has to be done using photographs of the metaphase spreads resulting in a loss of resolution. This does not occur with G-banding where the analysis is carried out directly under the microscope. Photographs can be taken for record purposes, although the slides should remain stable for a number of years. However Q-banding is particularly valuable for looking at the Y chromosome in man. With G-banding the Y chromosome appears uniformly grey and fairly nondescript, but with Q-banding it shows brilliant fluorescence and is very distinctive.

G11 banding. When examining human/rodent hybrids there are two particularly useful techniques available; G11 banding (Bobrow & Cross, 1974) and the more high technology method of *in situ* hybridisation (Mitchell *et al.*, 1986). G11 banding uses Giemsa at pH 11 instead of the usual pH 6.8 which results in differential staining of species. Mouse chromosomes stain dark magenta with pale blue centromeres, and human chromosomes stain blue with red differentiation of paracentric regions of chromosomes 1, 4, 5, 7, 9, 10, 13, 14, 15, 20, 21 and 22. *In situ* hybridisation is a technique which uses either radioactively or biotin labelled DNA probes to hybridise to metaphase chromosomes. This is a highly sensitive technique which enables the direct localisation of single copy genes on mammalian chromosomes.

A number of other techniques are known which highlight particular areas of the chromosome, e.g. silver staining (Bloom & Goodpasture, 1976) which stains nucleolar organising regions and C-banding (Pardue & Gall, 1970), which stains the constitutive heterochromatin of chromosomes.

Technical points regarding the technique. In order to achieve a good preparation for banding and analysis the cell line must be in the exponential phase of growth. Cell division is halted at metaphase by the addition of colcemid (Sigma, final concentration 0.1–0.4 $\mu g\ ml^{-1}$) for 1–6 h. Care should be taken not to extend the exposure time as this may give rise to chromosomes which are too short for satisfactory analysis. The incubation time for colcemid should reflect the division time of the cell line under consideration, e.g. a diploid Human cell line with a relatively long doubling time would generally require longer incubation than a rapidly proliferating Chinese Hamster ovary cell line. Cells are harvested and then exposed to a hypotonic solution (KCl) for up to 20 min and then fixed in methanol/acetic acid, 3:1 (v/v) at 4 °C. The length of time necessary in hypotonic solution does vary considerably between cell lines, e.g. Chinese Hamster cell lines require only 0.1 M KCl for 3 min to produce well spread chromosomes whereas Human fibroblasts require 10 min in 0.07 M KCl to achieve the same result.

Timing is not critical after the first fixation but insufficient fixation interferes with the flattening of the preparation onto the slide and coating substances can interfere with banding. The fixation step should be repeated until the cell pellet is white and the supernatant clear. This will normally take two changes of fresh cold fixative.

For slide making, the cell suspension should be centrifuged and resuspended in sufficient fresh fixative to give an opaque fluid. Slides can be made in a number of ways, e.g. using cold wet slides and drying on a hot plate. Some workers prefer to flame-dry slides while others air-dry. The technique of choice in our laboratory is to create a film of moisture simply by breathing on to the slides and then place a drop of cell suspension on to the slide which is held at an angle of 45 °C. The drop is allowed to run down the slide and is followed by an additional drop of fresh fixative. The slide is air-dried. The number of drops of fresh fixative required will vary according to the fragility of the cells. If it is difficult to obtain well spread chromosomes it is often easier to change the slide making technique before returning to a new culture and altering the harvest technique. Once made, the slides are examined using phase contrast microscopy and adjustments can be made to cell concentration and slide making accordingly.

Once the slides are prepared a suitable banding technique is chosen. For most banding techniques it is considered that slides which are 3–7 days old give best results, although by minor alterations to the methodology it is possible to band slides on the same day.

Further reading. For more detailed information on culturing and banding techniques, *Techniques in Somatic Cell Genetics* edited by Shay, has a chapter on chromosome analysis which as well as outlining the techniques also gives some technical notes on them. An excellent book for reference to techniques is *Human Cytogenetics, A Practical Approach,* edited by Rooney & Czepulkowski. Although the book gives protocols for use with human preparations they should give a good basis to build upon for other species and cell lines.

A number of reviews on the nature of banding are in the literature (Hsu, 1974; Sumner, 1982) and these may be useful in explaining some of the processes involved in banding chromosomes, although at present the mechanisms are not well understood.

When performing analysis it is always useful to have material to use as a reference and so for this purpose ideograms of normal karyotypes are used. Where human cell lines are involved the usual reference work is the ISCN, 1981, which is an international agreement on the

Fig. 5.3. DNA fingerprints of two Human T-cell lines, Jurkat (13822) and JM (13823). M tracks are standard references. Probe 33.15 shows virtual identity whereas the single locus probe mix shows genotypic differences.

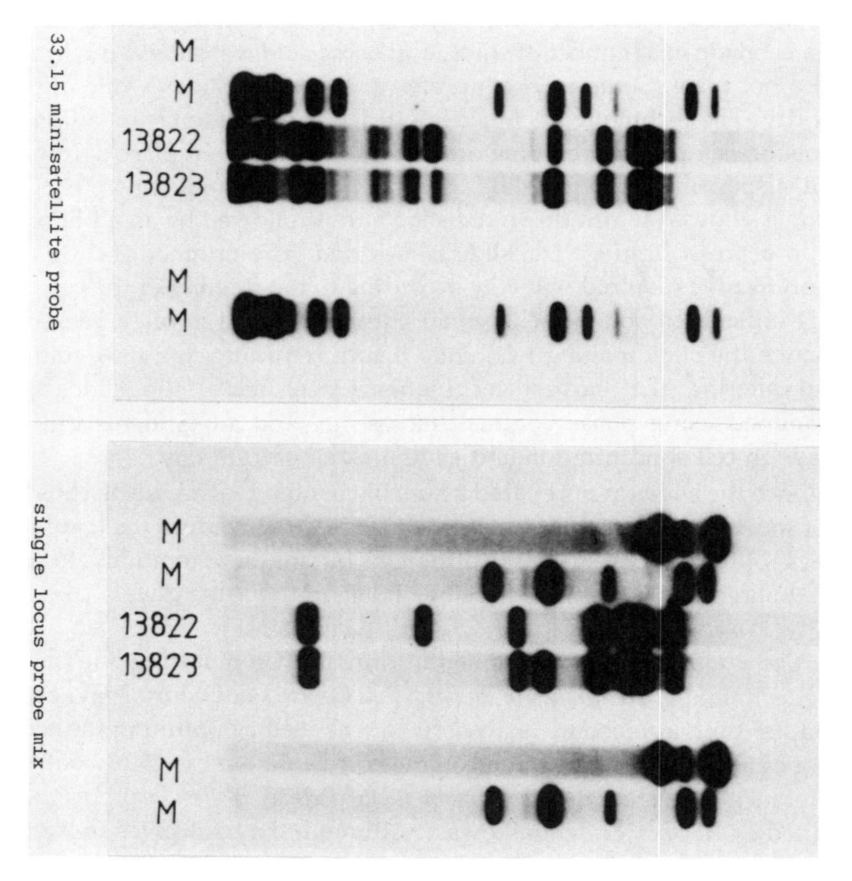

nomenclature for chromosome banding patterns. This type of agreement does exist for other species, notably the mouse.

5.4.3 *DNA fingerprinting*

A recent technique developed using DNA probes is DNA fingerprinting produced by restriction fragment length polymorphisms (RFLPs). Cellular DNA is digested with restriction enzymes, fractionated by agarose gel electrophoresis and the DNA transferred to a membrane by Southern blotting. The membranes are then exposed to radioactive minisatellite probes that react with complementary DNA fragments. These patterns are then developed on autoradiographs (Jeffreys *et al.*, 1985).

Figure 5.3 shows an example of this technique using Human T-cell lines JM and Jurkat, which were derived from the same individual but reveal genotypic differences with the single locus probe mix.

The technique is not only useful for human cells, it is possible to derive DNA fingerprints for all species (ECACC, unpublished data). It is possible that this method will become the standard reference technique for cell line identity.

5.5 References

Aula, P. & Nichols, W. W. (1967). The cytogenetic effects of mycoplasma in human leucocyte cultures. *Journal of Cell Physiology* **70**, 281–90.

Barile, M. F. (1977). Mycoplasma contamination of cell cultures: a status report. In *Cell Culture and its Application*, ed. R. T. Acton and J. D. Lynn, pp. 291–334. New York: Academic Press.

Barile, M. F. & Levinthal, B. G. (1968). Possible mechanism for mycoplasma inhibition of lymphocyte transformation induced by phytohaemagglutinin. *Nature* **219**, 751–2.

Bloom, S. E. & Goodpasture, C. (1976). An improved technique for selective silver staining of nucleolar organizer regions in human chromosomes. *Human Genetics* **34**, 199–206.

Bobrow, M. & Cross, J. (1974). Differential staining of human and mouse chromosomes in interspecific cell hybrids. *Nature* **251**, 77–9.

Butler, M. & Leach, R. H. (1964). A mycoplasma which induces acidity and cytopathic effect in tissue culture. *Journal of General Microbiology* **34**, 285–94.

Callewaert, D. M., Kaplan, J., Peterson, W. D. & Lightbody, J. W. (1975). Suppression of lymphocyte activation by a factor produced by *Mycoplasma arginini*. *Journal of Immunology* **115**, 1662–6.

Caspersson, T., Lomakka, G. & Zeck, L. (1971). The 24 fluorescence patterns of the human metaphase chromosomes – distinguishing characters and variability. *Heridatas* **567**, 89–102.

Chen, T. R. (1977). *In situ* demonstration of mycoplasma contamination in cell cultures by fluorescent Hoechst 33258 stain. *Experimental Cell Research* **104**, 255–62.

Commission of the European Communities (1988). Guidelines on the production and quality control of monoclonal antibodies of murine origin intended for use in man. *Trends in Biotechnology* **6**, G5–G8.

Committee on Standardized Genetic Nomenclature for Mice: Standard karyotype of the mouse, *Mus musculus* 1972. *Journal of Heredity* **63**, 69–72.

Esber, E. C. (1987). Points to consider in the characterisation of cell lines used to produce biologicals: Dept. of Health and Human Series, US Food and Drug Administration, Bethesda, MD.

European Pharmacopoeia (1980). *Biological Tests*, 2nd edn, Part 1, Vol. 2. France: Maisonneuve, S.A.

Fogh, J. & Fogh, H. (1967). Morphological and quantitative aspects of mycoplasma–human cell relationships. *Proceedings of the Society for Experimental Biology and Medicine* **125**, 423–30.

Halton, D. M., Peterson, W. D., Jr & Hukku, B. (1983). Cell culture quality control by rapid isoenzymatic characterisation. *In Vitro* **19**, 16–24.

Harris, H. & Hopkinson, D. A. (1976). *Handbook of Enzyme Electrophoresis in Human Genetics*. Amsterdam: North-Holland Publishing Co.

Hessling, J. J., Miller, S. E. & Levy, N. L. (1980). A direct comparison of procedures for the detection of mycoplasma in tissue culture. *Journal of Immunological Methods* **38**, 3154–324.

Hopps, H. E., Meyer, B.C., Barile, M. F. & DelGiudice, R. A. (1973). Problems concerning non-cultivable mycoplasma contamination in tissue cultures. *Annals of The New York Academy of Sciences* **225**, 265–76.

Hsu, T. C. (1974). Longitudinal differentiation of chromosomes. *Annual Review of Genetics* **7**, 153–76.

ISCN (1981). An International System for Human Cytogenetic Nomenclature – High resolution banding birth defects: Original Article series, Vol. XVII, No. 5. New York: March of Dimes Birth Defects Foundation.

Jeffreys, A. J., Wilson, V. & Thein, S. L. (1985). Hyperviable 'minisatellite' regions in human DNA. *Nature* **314**, 67–73.

Kotani, H., Phillips, D. & McGarrity, G. J. (1986). Malignant transformation of NIH-3T3 and CV-1 Cells by a helical mycoplasma, *Spiroplasma mirum*, strain SMCA. *In Vitro Cellular and Developmental Biology* **22**, 756–62.

Kraemer, P. M., Defendi, V., Hayflick, L. & Manson, L. (1963). Mycoplasma (PPLO) strains with lytic activity for murine lymphoma cells *in vitro*. *Proceedings of the Society for Experimental Botany and Medicine* **112**, 281–387.

Levine, E. M., Thomas, L., McGregor, D., Hayflick, L. V. & Eagle, H. (1968). Altered nucleic acid metabolism in human cell cultures infected with mycoplasma. *Proceedings of the National Academy of Sciences, Washington* **60**, 583–9.

MacPherson, I. & Russell, W. (1966). Transformation in hamster cells mediated by mycoplasmas. *Nature* **210**, 1343–5.

Marcus, M., Lavi, U., Nattenberg, A., Rottem, S. & Markowitz, O. (1980). Selective killing of mycoplasmas from contaminated mammalian cells in cell cultures. *Nature* **285**, 659–61.

McGarrity, G. J. (1982). Detection of mycoplasmal infection of cell cultures. In *Advances in Cell Cultures*, ed. K. Maramorosch, pp. 99–131. New York: Academic Press.

McGarrity, G. J. & Coriell, L. L. (1973). Detection of anaerobic mycoplasma in cell cultures. *In Vitro* **8**, 17–18.

McGarrity, G. J., Kotani, H. & Carson, D. (1986). Comparative studies to determine the efficiency of 6-methylpurine deoxyriboside to detect cell culture mycoplasmas. *In Vitro Cellular and Developmental Biology* **22**, 301–4.

McGarrity, G. J., Sarama, J. & Vanaman, V. (1979). Factors influencing microbiological detection of mycoplasmas in cell culture. *In Vitro* **15**, 73–81.

McGarrity, G. J., Vanaman, V. & Saraman, J. (1978). Methods of prevention, control and elimination of mycoplasmal infection in *Mycoplasma Infection of Cell Cultures*, ed. G. J. McGarrity, D. G. Murphy and W. W. Nichols, pp. 213–41. New York: Plenum Press.

Mitchell, A. R., Ambros, P., Gosden, J. R., Morten, J. E. N. & Portius, D. J. (1986). Gene mapping and physical arrangements of human chromatin in transformed hybrid cells: fluorescent and autoradiographic *in situ* hybridization compared. *Somatic Cell and Molecular Genetics* **12**, 313–24.

Mowles, J. M. (1988). The use of Ciprofloxacin for the elimination of mycoplasma from naturally infected cell lines. *Cytotechnology* **1**, 355–8.

Nair, C. N. (1985). Elimination of mycoplasma contaminants from cell cultures with animal serum. *Proceedings of the Society for Experimental Biology and Medicine* **179**, 254–8.

Neftel, K. A., Müller, M. R., Widmer, U. & Hügin, A. W. (1986). β-lactam antibiotics inhibit human *in vitro* granulopoeisis and proliferation of some other cell types. *Cell Biology and Toxicology* **1**, 513–21.

Nelson-Rees, W. A., Daniels, D. W. & Flandermeyer, R. R. (1981). Cross-contamination of cells in culture. *Science* **212**, 446–52.

Pardue, M. L. & Gall, J. G. (1970). Chromosomal localization of mouse satellite DNA. *Science* **168**, 1356–8.

Paton, G. R., Jacobs, J. P. & Perkins, F. T. (1965). Chromosome changes in human diploid cell cultures infected with mycoplasma. *Nature* **207**, 43–5.

Razin, S. (1983). Identification of mycoplasma colonies. In *Methods in Mycoplasmology*, Vol. 1, ed. S. Razin and J. G. Tully, pp. 83–8. New York: Academic Press.

Robinson, L. B., Wichelhausen, R. B. & Roizman, B. (1956). Contamination of human cell cultures by *Pleuropneumonia*-like organisms. *Science* **124**, 1147–8.

Rooney, D. E. & Czepulkowski, B. H. (1986). *Human Cytogenetics*. Oxford: IRL Press.

Seabright, M. (1972). A rapid banding technique for human chromosomes. *Lancet* **ii**, 971–2.

Schmidt, J. & Erfle, V. (1984). Elimination of mycoplasma from cell cultures and establishment of mycoplasma-free cell lines. *Experimental Cell Research* **152**, 565–70.

Shay, J. W. (1982). *Techniques in Somatic Cell Genetics*. New York: Plenum Press.

Simberkoff, M. S., Thorbecke, G. J. & Thomas, L. (1969). Studies of PPLO infection. V. Inhibitors of lymphocyte mitosis and antibody formation by mycoplasmal extracts. *Journal of Experimental Medicine* **129**, 1163–81.

Stanbridge, E. J., Hayflick, L. & Perkins, F. T. (1971). Modification of amino-acid concentrations induced by mycoplasmas in cell culture medium. *Nature New Biology* **232**, 242–4.

Sumner, A. T. (1982). The nature of chromosome bands and their significance for cancer research. *Anticancer Research* **1**, 205–16.

Sumner, A. T., Evans, H. J. & Buckland, R. A. (1971). New technique for distinguishing between human chromosomes. *Nature New Biology* **232**, 31–2.

Wise, K. S., Cassell, G. H. & Acton, R. T. (1978). Selective association of murine T-lymphoblastoid cell surface alloantigens with *Mycoplasma hyorhinis*. *Proceedings of the National Academy of Sciences, Washington* **75**, 4479–83.

Van Diggelen, O., Shin, S. & Phillips, D. (1977). Reduction in cellular tumorigenicity after mycoplasma infection and elimination of mycoplasma from infected cultures by passage in nude mice. *Cancer Research* **37**, 2680–7.

United States Pharmacopeia (1985). *Sterility Tests*. 21st Revision, pp. 1156–60. United States Pharmacopeial Convention, Inc.

6

Patent protection for biotechnological inventions

I. J. BOUSFIELD

6.1 Introduction

This chapter is intended to give the reader who is unfamiliar with patents an introduction to the patent system as it applies to biotechnology, and a general guide to the procedures and pitfalls involved in obtaining patent protection for biotechnological inventions. For a detailed discussion of the whole subject of patents in biotechnology and a review of the variety of national patent systems the reader is referred to the excellent texts by Crespi (1982), Crespi & Straus (1985) and Straus (1985). It is not possible here to provide a step-by-step guide to getting a patent in every country in the world, for, despite an overall similarity, variations between different national patent laws are manifold, and professional help is necessary to guide even the experienced inventor through their complexities. The present account does no more than skim the surface of what is a complex and often fascinating subject; for this reason a short list of selected publications which illustrate in more detail many of the points raised here is given in Section 6.6, Further reading.

6.2 Basis of the patent system

6.2.1 *Principles*

The principle (if not the practice) of the patent system is straightforward: the inventor of a new product or process publicly discloses the details of his invention and in return he is granted for a limited period a legally enforceable right to exclude others from exploiting it. In this way the inventor's ingenuity is acknowledged and rewarded, while at the same time further technical progress is encouraged by the public dissemination of information about the invention.

6.2.2 *Criteria for patentability*

To qualify for patent protection, an invention must meet the following major criteria.

Novelty. The invention must be *new*. Most countries apply the test of 'absolute novelty', which means that if prior knowledge of it exists anywhere in the world (not merely in the country where patent protection is sought) then the invention belongs to the state of the art ('prior art') and is not patentable. The prior art is held in these countries to include anything the inventor himself may have said or written about his invention. Exceptions to this rule of absolute novelty are the patent systems of the USA and Canada, where publications by the inventor made not more than one (USA) or two (Canada) years before a patent application is filed in that country do not destroy novelty. 'Grace periods' of six months are also allowed by Australia, New Zealand and Japan, but only in respect of certain kinds of publications made by the inventor, for example at certain scientific meetings. Under nearly all patent systems, publications made after the date that the patent application is filed (the 'priority date') do not jeopardise protection for that particular subject matter in that particular application. It must be remembered, however, that they will form part of the prior art against which any *future* applications will be assessed.

Inventiveness. The invention must show evidence of an *inventive step*, that is it must not be 'obvious' from the state of the art to anyone 'skilled in the art'. In simple terms this means that the average expert in the field under consideration could not reasonably have predicted the invention as an obvious or logical outcome of what he already knew.

Utility (industrial applicability). The invention must have a *practical use*, which in the USA means exactly that, but in nearly all other patent systems means that it must also be capable of industrial application. Most countries, however, hold that medical methods for the direct treatment of the human or animal body are not susceptible to industrial application and are therefore not patentable. The major exception to this is again the USA.

Disclosure. The details of the invention must be *disclosed* by means of a patent specification (see Section 6.5.2 below) so that the invention is described in sufficient detail to allow a skilled person to reproduce it. This is normally done by means of a written description supplemented where necessary by drawings. However, in one major category of biotechnological inventions – those involving the use of new microorganisms and certain other novel living materials – a written description is not usually considered sufficient for the purposes of disclosure. In such cases it is argued that no matter how carefully the description may be worded, if the microorganism itself is not available, the invention cannot be reproduced. Therefore, many countries require the deposit of new microorganisms in a recognised culture collection to ensure their public availability. This unique aspect of biotechnological patent procedure is dealt with in some detail later in this chapter.

Exclusions for patentability. As well as meeting the criteria listed above, an invention must not be of a kind which is excluded from patentability by its very nature. Exclusions relating specifically to biotechnological inventions are discussed later, but in general terms patents cannot be obtained for mere discoveries, theories, computer programs, literary works, musical compositions, aesthetic creations and illegal or offensive devices.

6.3 Kinds of biotechnological invention

There are four main kinds of biotechnological invention: products, compositions, processes, and use or methods of use (Crespi, 1982).

6.3.1 *Products*

These inventions are exactly what the word suggests – tangible new materials or entities. They include organisms themselves (e.g. bacteria, fungi), parts of organisms (e.g. cell lines), substances produced by either of these (e.g. enzymes, antibiotics), and substances obtained by or employed in recombinant DNA techniques (e.g. plasmids, DNA molecules).

Product inventions can be the subject of two broad kinds of patent claim: the 'product *per se*', where patent protection is sought for the product itself, regardless of the method of manufacture, and the more limited 'product-by-process', where protection is sought for the product obtained by a particular process.

6.3.2 *Compositions*

These inventions are mixtures of substances or organisms, the individual components of which may already be known, but which in combination can be shown to display a new property or exert a new effect.

6.3.3 *Processes*

These inventions are methods for the manufacture of products, and include bioconversions, fermentations, and methods of isolation, purification or cultivation. Some process inventions are genuinely new methods for obtaining novel or known products, but others are known methods applied in new situations or used in the manufacture of novel products.

6.3.4 *Use and methods of use*

Methods of use include processing or treating materials (e.g. industrial raw materials or agricultural products), non-medical treatments of humans or animals, 'off the body' medical methods (e.g. a method of diagnosis carried out on a sample taken from a patient), methods of testing (e.g. quality control) and in a few countries, notably the USA, medical treatment of humans and animals. Also, the new medical use of a substance previously unknown to have that use is protectable in European Patent Convention (EPC) countries.

6.4 Patentability of biotechnological inventions

6.4.1 *Inventions involving new plants and animals*

Plant varieties. By far the most common form of legal protection for new plant varieties is the plant variety right (although in the USA special 'plant patents' are available for asexually propagated plants). Several countries are now party to the International Convention for the Protection of New Varieties of Plants (UPOV) which aims to harmonise national practices as far as possible (International Convention, 1978). In these countries, plants that are protected by plant variety rights are usually specifically excluded from patentability. Plant variety rights in general are intended to allow the commercial plant breeder a monopoly on the production of propagating material for the purposes of commercial marketing, its offer for sale and its marketing. Plant variety rights are easier to obtain than is patent protection as there is no requirement for inventiveness or for reproducible disclosure, but they are more limited in scope in that neither the plant itself nor consumables produced from it (e.g. fruit for eating, grain for milling) are protected.

Plant variety rights were introduced essentially to cover varieties developed by traditional breeding methods and it is such varieties that are excluded from patentability in many countries. Thus Article 53(b) of the EPC, to which 13 countries belong (Table 6.1) states the following:

> European patents shall not be granted in respect of . . . plant or animal varieties or essentially biological processes for the production of plants or animals; this provision does not apply to microbiological processes or the products thereof.

The same exclusion is found where the national laws of individual European countries have been harmonised with the EPC and is also contained in the laws of several other countries, e.g. the German Democratic Republic, Mexico, Sri Lanka, Thailand and Yugoslavia. Plant and animal varieties, but not essentially biological processes, are also excluded from patentability in China.

The exclusion of plant varieties from patent protection is a contentious issue. Straus (1985) has pointed out that current systems for the protection of plant varieties were introduced when plant breeding methods did not permit the breeder to fulfil the normal criteria of patentability. However, the advent of new technologies, particularly genetic manipulation techniques, for the production of new plant varieties has meant that these requirements can now be met. Therefore, there seems to be a good argument in favour of allowing the developer of such varieties the right to choose between protection

Table 6.1. *Countries party to the European Patent Convention at 1 January 1987*

Austria	Luxembourg
Belgium	Netherlands
France	Spain
Germany (Federal Republic)	Sweden
Greece	Switzerland
Italy	United Kingdom
Liechtenstein	

under the patent system or through plant variety rights (Beier *et al.*, 1985; Straus, 1985).

Animal varieties. Although a special form of protection for new animal varieties (which are not, however, regarded as inventions) is available in the Soviet Union, there is in general no special system of legislation for their protection. Most of the countries that exclude plant varieties from patentability also exclude animal varieties, and in those countries that do not, the position is not altogether clear. However, recent court decisions in the USA (*in re* Diamond & Chakrabarty, 1980) and Canada (*in re* Abitibi, 1982) suggest that animals may be patentable provided that the requirements for enabling disclosure, that is repeatability, are met (Straus 1985).

Processes for the production of animals and plants. Many patent laws, including those of the EPC, specifically deny patent protection for 'essentially biological processes' for the production of plants or animals; microbiological processes, however, are not included in this provision. This terminology may not be entirely clear and perhaps needs some explanation. Stated simply, an essentially biological process is considered to be one in which the result is achieved with a minimum of human technical intervention. The example given by the European Patent Office (EPO) in its guidelines to examiners is a method for selectively breeding horses in which human intervention is limited to bringing together animals having particular characteristics. On the other hand, a process for treating a plant to promote or suppress its growth, e.g. a method of pruning or of applying stimulatory or inhibitory substances, would not be considered to be essentially biological, since although a biological process is involved, the essence of the invention is technical.

Given this definition of 'essentially biological' and the exemption of microbiological processes from it, the normal criteria for patentability can be applied to methods for producing plants by, for instance, genetic manipulation involving the use of vectors in microbial hosts, or by somatic cell hybridisation. However, matters are less certain under the EPC in respect of some processes for the production of new animal varieties, even though such processes may meet the test of not being essentially biological. This is because a further exclusion from patentability is found in Article 52(4) of the EPC, which states the following:

> Methods for treatment of the human or animal body by surgery or therapy and diagnostic methods practised on the human or animal body shall not be regarded as inventions which are susceptible of industrial application . . . [See also Utility in Section 6.2.2 above.]

Straus (1985) has expressed the fear that some present and future approaches to animal breeding, such as techniques of embryo transfer, could be denied patent protection by this provision. In support of his argument he cites a recent decision by the UK Comptroller of Patents, in which an application involving just such a technique was rejected as being a method of treatment by surgery.

In contrast to the European system, the patent laws of the USA, Japan and China do not exclude essentially biological processes. Furthermore, the US laws do not exclude methods for treating animals (or humans); therefore the problems presented by the European system in respect of patenting processes involved in animal breeding do not exist in the USA.

Tissue cultures. Animal cell lines and plant tissue cultures (and in Japan, seeds) are generally considered to be in the same category as microorganisms for patent purposes. Thus they are subject to the provisions applied to microbiological inventions as discussed below. As regards plant cells, however, the US Patent Office makes a distinction between undifferentiated cell lines used, for instance, to produce a particular substance, and cells which are capable of differentiation and which are used simply to reproduce the whole plant.

6.4.2 *Inventions involving microorganisms*

Applied microbiology in its broadest sense is a major facet of modern biotechnology, and any discussion of patents in biotechnology inevitably must focus on the peculiar problems posed by microbiological inventions. In fact so great has been the attention given to these problems in patent circles that the patent legislation of an increasing number of countries contains specific provisions for inventions involving the use of microorganisms, and one international convention (the Budapest Treaty; see below) deals entirely with microorganisms (Budapest Treaty and Regulations, 1982).

It should be said that the term 'microorganism' is used in patent circles in a very loose sense and encompasses entities that strictly speaking are not microorganisms, e.g. cell lines and plasmids. Indeed, the word is intentionally not defined in the Budapest Treaty so as to avoid undue constraints being imposed upon the application of the Treaty, and in the words of the World Intellectual Property Organisation (WIPO) commentary on the draft Treaty (WIPO, 1980), it 'need not correspond to usage in some scientific circles'. Unfortunately, the use of such inexact terminology has led to uncertainty in some quarters as to what is or is not a microorganism. Because of this, the present author (acting on behalf of the World Federation for Culture Collections (WFCC) Patents Committee) has proposed to WIPO that the expression 'living material' be used instead of the word microorganism, particularly in regard to the Budapest Treaty. The word 'living' was defined as 'that material which under appropriate conditions is able to replicate itself, or which at least possesses the functional genes necessary to direct its own replication'. This definition has two advantages: first, it avoids insoluble philosophical arguments about where chemical reactivity ends and life begins and, second, it excludes non-living biological materials such as enzymes. In the present chapter, although the word microorganism is used for ease of reference, it should be taken to mean living material as defined above.

Microbiological inventions may be found in all of the categories of biotechnological invention outlined in Section 6.3 above. In general, microbiological processes and the (inanimate) products obtained by them can be considered analogous to chemical processes and products, and obtaining patent protection for them in most countries is nowadays fairly straightforward, provided that the basic criteria for patentability are fulfilled. Less straightforward, however, is the patenting of microorganisms as products either *per se* or as products-by-process. It is these inventions above all others that demonstrate the difficulties of determining the borderline between 'discoveries' and 'inventions', of what is 'new' and what is not, and of ensuring sufficiency of disclosure.

Naturally occurring microorganisms. A previously unknown naturally occurring microorganism that is left in its natural state is universally regarded as a discovery and is unpatentable. However, the degree of human intervention considered necessary to turn such a discovery into a patentable invention (assuming it has a practical use) varies between different countries. The extent of this variation was demonstrated by the official replies to a questionnaire on patent protection in biotechnology distributed to governments in 1982 by the Organization for Economic Cooperation and Development (OECD). These responses were reviewed in detail by Crespi (1985). In those countries that do permit naturally occurring organisms to be patented, isolation and purification of the organism are general prerequisites, after which various constraints are applied, mainly relating to novelty and unexpected properties. Thus, for example, the UK and the EPO require an organism to be 'new' in the sense of being hitherto *unknown*, whereas in Canada a new organism is one that does not already *exist* in nature (in this connection, Crespi (1985) has commented pointedly on the illogicality of equating 'unknown' with 'not previously existing'); the Federal Republic of Germany requires that 'certain changes occur during isolation so that the isolated microorganism is not identical with that occurring in nature'; Denmark requires naturally occurring organisms to have unforeseen properties. The USA permits the patenting of naturally occurring organisms as 'biologically pure cultures'.

Non-naturally occurring microorganisms. After the much-publicised Chakrabarty case in the USA in 1980, in which a genetically manipulated strain of *Pseudomonas* was held not to be a product of nature but a human invention patentable *per se*, there are unlikely to be any unusual problems in obtaining patent protection for 'artificial' microorganisms (bacterial recombinants, hybridomas, etc.), other than in countries which do not permit the patenting of any kinds of microorganism.

Sufficiency of disclosure. As already mentioned (Disclosure, in Section 6.2.2 above), one of the fundamental requirements of the patent system is that the details of an invention must be disclosed in a manner sufficient to allow a skilled person to reproduce the invention. Microbiological inventions present particular problems of disclosure in that more often than not repeatability cannot be ensured by means of a written description alone. In the case of an organism isolated from soil, for instance, and perhaps 'improved' by mutation and further selection, it would be virtually impossible to describe the strain and its selection sufficiently to guarantee another person obtaining the same strain from soil himself. In such a case, the strain itself forms an essential part of the disclosure. In view of this an increasing number of countries require a 'new' microorganism (i.e. one not already generally available to the public) to be deposited in a recognised culture collection whence it can subsequently be made available at some stage in the patent procedure. As a general principle, an invention should be reproducible from its description at the time that the patent application is filed. In the case of an invention involving a new organism, therefore, most patent offices require a culture of the organism to be deposited not later than the filing date of the application (or the priority date if priority is claimed from an earlier application – see Section 6.5.3 below). Exactly when a strain becomes available varies according to the patent laws of different countries, and is a much debated question dealt with in more detail below (Release of samples). Since an invention must also be reproducible throughout the life of the patent, a microorganism deposited for patent purposes must remain available for at least this length of time. Most countries provide for a considerable safety margin in this respect, and availability for at least 30 years is a common requirement.

The Budapest Treaty. In order to obviate the need for inventors to deposit their organism in a culture collection in every country in which they intend to seek patent protection, the 'Budapest Treaty on the International Recognition of the Deposit of Microorganisms for the Purposes of Patent Procedure' was concluded in 1977 and came into force towards the end of 1980 (Budapest Treaty and Regulations, 1981). Under the Budapest Treaty certain culture collections are recognised as 'International Depositary Authorities' (IDAs), and a single deposit made in any one of them is acceptable by each country party to the Treaty as meeting the deposit requirements of its own national laws. Any culture collection can become an IDA provided that it has been formally nominated by a contracting state, which must also provide assurances that the collection can comply with the requirements of the Treaty. At 30 November 1989 there were 20 IDAs and they and the kinds of organisms they accept are listed in Table 6.5 below.

The Budapest Treaty provides an internationally uniform system of deposit and lays down the procedures which depositor and depository must follow (see Section 6.5.5 below), the duration of deposit (at least 30 years or 5 years after the most recent request for a culture, whichever is later), and the mechanisms for the release of samples. The Treaty does not, however, concern itself with the *timing* of deposit nor, in the main, of release; these are determined by the relevant national laws. Likewise, the recipients of samples (other than patent offices and people with the depositor's authorisation) are referred to merely as 'parties legally entitled': exactly *who* such parties are and under what conditions they may obtain samples are again determined by national law. Twenty one states and the European Patent Office (EPO) are now party to the Budapest Treaty and are listed in Table 6.2.

National deposit requirements. The requirements of various countries for the deposit and release of microorganisms for patent purposes are summarised in Table 6.3. Deposit is a statutory requirement under Rule 28 of the EPC and under the national laws of several of its member countries. In those EPC countries not having a statutory provision under their national law, deposit is such an established requirement of patent offices that it amounts to the same thing for all practical purposes. Most of these countries follow EPC practice in requiring the deposit to be made by the filing or priority date. An exception to this is the Netherlands, where deposit is required before the second publication (see next section) of the patent application. Most other European countries do not have specific requirements as yet, but nevertheless advise that deposits should be made, usually along the lines of the EPC.

In many countries outside Europe, deposit is an established or recommended practice, and some patent offices (in Japan, USA, USSR, for example) have specific requirements for deposit. In almost all cases deposit must be made by the filing or priority date. In the USA,

Table 6.2. *Countries party to the Budapest Treaty at 31 July 1987*

Australia	Liechtenstein
Austria	Netherlands
Belgium	Norway
Bulgaria	Philippines
Denmark	South Korea
Finland	Spain
France	Sweden
German Democratic Republic	Switzerland
Federal Republic of Germany	UK
Italy	USA
Hungary	USSR
Japan	European Patent Office (EPO)[a]

[a] The EPO is not, strictly speaking, a party to the Treaty, since it is not a country but an intergovernmental organisation. Article 9 of the Treaty provides for such organisations to file a declaration stating that they accept the obligations and provisions of the Treaty. The EPO has filed such a declaration.

Table 6.3. *National requirements (mandatory or recommended) for deposit and release of microorganisms**

Country	Deposit by	Earliest release	Earliest general availability[a]	Restrictions on distribution and use of samples	Minimum storage period (years)
Australia	F/P	1st pub.	1st pub.	as for UK	30
Austria	F/P	–	–	–	–
Belgium	F/P	1st pub.	grant	as for EPO	30
Bulgaria	F/P	grant	grant	as for UK	–
Canada	F/P	grant	grant	none	life of patent
Denmark	F/P	1st pub.	grant	as for EPO	30
Finland	F/P	1st pub.	grant	as for EPO	30
France	F/P	1st pub.	grant	as for EPO	30
Germany	F/P	1st pub.	1st pub.	until patent expires, sample must not be passed to 3rd parties or outside purview of German law	20
Hungary	F/P	1st pub.	1st pub.	sample must not be passed to 3rd party	20
Ireland	F/P	–	–	–	–
Italy	F/P	?	?	as for EPO	?
Japan	F/P	2nd pub.	2nd pub.	sample must not be passed to third parties until patent expires and must be used only for research purposes	life of patent
Liechtenstein				as for Switzerland	

Netherlands	2nd pub.	2nd pub.	2nd pub.	none	life of patent
New Zealand	F/P	–	–	–	–
Norway	F/P	1st pub.	grant	as for EPO	30
Portugal	F/P	–	–	–	–
Spain	F/P	1st pub.	1st pub.	–	–
Sweden	F/P	1st pub.	grant	as for EPO	30
Switzerland	F/P	?	?	sample must not be passed to 3rd parties	30
UK	F/P	1st pub.	1st pub.	sample must not be passed to 3rd parties until patent expires and must be used only for experimental purposes	30
USA[b]	F/P	grant	grant	none	30
USSR	–	–	–	–	–
EPO	F/P	1st pub.	grant	if applicant chooses, available only to independent expert before grant; must not be passed to 3rd parties before patent expires and must be used only for experimental purposes	30

F/P, filing or priority date as applicable; pub., pubication of application; –, no provisions or provisions not known; ?, conflicting information from different sources.

*Author should be contacted for information subsequent to publication of the first books in this series (Filamentous Fungi, Yeasts) in 1988.

[a] General availability means sample publicly available at least in country where application has been filed.

[b] The Lundak decision (1985) may mean that in certain cases deposit may be made later.

however, as a consequence of a recent court decision (*in re* Lundak, 1985), deposit may in certain circumstances be made after filing but before the issuance of a US patent.

All countries party to the Budapest Treaty (Table 6.2) must recognise a deposit made in an IDA but not all *require* deposits to be made in IDAs. Thus, for example, France, Germany, Switzerland, the UK, the USA and the EPC will recognise other culture collections that can comply with their particular requirements. Hungary accepts deposits made in collections on its own soil, but the only deposits it will recognise elsewhere are those made in IDAs. The Japanese patent office, however, will recognise deposits outside Japan only if they have been made under the Budapest Treaty or have been 'converted' to Budapest Treaty deposits (see 'Converted' deposits in Section 6.5.5 below), regardless of their previous public availability. It must be remembered that deposits made under the Budapest Treaty can only be made in IDAs.

Most countries not party to the Budapest Treaty accept deposits made in any internationally known culture collection which will comply with their requirements; in some cases the collection is required to furnish a declaration as to the permanence and availability of the deposit.

Release of samples. Microorganisms deposited to comply with requirements for disclosure must become available to the public at some stage of the patenting procedure. Unlike a written description, however, the microorganism is the physical essence of the invention itself and because of this the exact conditions of release are a matter of great concern to patent practitioners. There are as yet no internationally uniform release conditions, but three main kinds of system operate at present.

In the USA, patent applications are not published until the patent is granted, and a microorganism deposited for patent purposes need not be made available until then. From the date of grant, the organism must be publicly available without any restriction. The major advantage of this system to the inventor is that his microorganism does not have to be released until he has an enforceable right. Furthermore, if he is *not* granted a patent then his microorganism need never become available. Thus under the US system an inventor is never put in the position of having to allow access to his organism when he has no legal protection.

In Japan and the Netherlands patent applications are published twice, first 18 months after the filing date of priority date and before the application has been examined (Section 6.5.4 below), and second (for the purposes of opposition of third parties) when the application has been accepted. A microorganism deposited in connection with the application must be made available at the date of second publication, i.e. once the patent office has decided to grant a patent. Thus under the Japanese and Dutch systems the inventor again has an enforceable right at the time he is required to make his organism available. In Japan he is afforded a further measure of protection in that recipients of cultures must not pass them on to third parties and must use them only for experimental purposes. This provision does not apply under Dutch law, however.

A dual publication system is also operated by the EPC and by the countries party to it. However, in contrast to the Japanese requirements, a microorganism deposited in connection with an EPC application must be made available at the date of *first* publication, i.e. before any enforceable right exists. This practice reflects the prevailing philosophy of European patent authorities that the organism is regarded as an integral part of the disclosure and therefore should become available at the same time as the written description. Originally cultures had to be available to anyone requesting them, subject to the recipient giving certain undertakings of rather doubtful value, but in response to pressure from users of the system the appropriate rule (Rule 28) of the EPC was amended to provide more protection for the inventor. Rule 28(4) now permits the applicant to opt to restrict the availability of his organism at first publication to an independent expert acting on behalf of a third party. The expert, who is chosen by the requesting party from a list held by the EPO, is not permitted to pass cultures of the strain to anyone else. After second publication, the strain becomes generally available, but at this stage the inventor has an enforceable right.

Although this so-called 'expert solution' applies in respect of applications filed with the EPO itself, it is at present part of the national law of only a minority of member countries of the EPC (France, Italy, Sweden). None, however, permits recipients of cultures to pass them on to third parties.

Need for deposit. In principle most countries require deposit only when repeatability of the invention cannot be ensured without it. Thus, for example, it should not be necessary to deposit a new recombinant strain if the procedure for constructing the novel plasmid and transforming it into a host can be described in sufficient detail to allow an expert to produce the same recombinant for himself (given, of course, that the original vector and host are already generally available). In practice, however, applicants in such cases sometimes choose to deposit in order to avoid all risk of their application being rejected on the grounds of insufficient disclosure. Some applicants, on the other hand, prefer to take this risk.

In cases where the microorganism is already generally available from a culture collection, the situation is perhaps more straightforward. Some countries (e.g. Germany, USA, USSR) require the applicant to furnish a declaration signed by the culture collection and stating that the organism in question is in fact available and will remain available for the period dictated by the relevant national law. In this connection, it is worth noting that the USA in some cases presently requires such a declaration – at least for deposits made outside the USA – even where an organism has been deposited under the Budapest Treaty. As mentioned earlier, the Japanese patent office will recognise the availability of strains from culture collections outside Japan only if they have been deposited under the Budapest Treaty.

6.5 Practical considerations

So far this chapter has been concerned with the principles of biotechnological patents and the requirements of various countries; now the essentially practical aspects of the patenting process must be considered. To use specific examples, either actual or hypothetical, for this purpose would give too narrow a picture. Therefore, the more general question of seeking patent protection for an unspecified invention involving the use of a new (i.e. not already generally available) microorganism will be considered.

6.5.1 *The patent agent (patent attorney)*

The job of the patent agent is, put simply, to obtain a patent on behalf of the applicant. The agent's knowledge and understanding of patent procedures world-wide are essential to guide the applicant through the complex business of seeking patent protection, helping to draft the technical description, formulating the claims, dealing with the patent authorities, ensuring deadlines are met and so on. However, his technical knowledge of the invention and its background cannot be expected to equal that of the inventor, who must therefore be prepared to spend time and effort in familiarising him with every aspect. In his turn, the agent can often offer valuable advice on areas where more experimental work might be done before filing in order to make the application as strong as possible. The importance of these considerations is shown by the fact that many large firms have full-time patent agents on their staff to ensure that their inventions are adequately protected. The small organisation and the academic inventor, therefore, are well advised to obtain the services of a professionally qualified patent agent if they are considering applying for patent protection.

6.5.2 *Disclosing the invention*

Premature disclosure. As mentioned earlier, making a full disclosure of an invention is the applicant's side of the bargain that will give him a legal monopoly, and is a fundamental prerequisite for obtaining a patent. However, the disclosure must be made in the proper way and at the right time. Above all the invention must not be disclosed prematurely, for its novelty (see Section 6.2.2 above) will be assessed by most patent offices in the light of what is already known (the state of the art) on the day the patent application is filed. The prevailing state of the art includes any contributions the applicant himself may have made to it, whether orally, by visual display, or by display or sale of a product. Thus to avoid premature disclosure, the academic inventor in particular must abjure the normal practices of discussing his findings with other workers or publishing them in scientific journals until he has filed his patent application. Information revealed by breach of the applicant's confidence does not jeopardise a patent application, but since breach of confidence is often difficult to prove, the wisest course is to make no disclosure until the application has been safely filed. These strictures do not wholly apply (the exact conditions vary) in relation to those few countries, notably the USA, that allow a 'grace period' (see Section 6.2.2 above). However, even here it must be remembered that although the relevant national law may allow disclosure during a grace period preceding the basic national filing, such a disclosure will be prejudicial to subsequent foreign filings.

The patent specification – technical description. The patent specification contains the written disclosure or technical description of the invention and the patent claims, which state the scope and kinds of monopoly being asked for. The precise wording of the patent specification is of great importance and it is here that the skill of the patent agent comes into play in ensuring that the description (supplemented by deposit – see Section 6.5.5 below) fulfils the requirements of disclosure and that the claims are drafted to afford the best protection.

The technical description is exactly that; the invention is described in detail in scientific and technical terms – it is not addressed to the layman – and put into the context of the field to which it applies, the problems it aims to solve and the way in which solutions are achieved. The preferred format of the description is well established and typically includes the following: field, background, object and summary of the invention, followed by the detailed description of the invention. Crespi (1982) has discussed the layout of the technical description more fully, with actual examples, and the reader is referred to his book for more information. As far as the present account is concerned, the first point is that the description must clearly convey the novelty, inventiveness and industrial applicability of the invention, and must describe the methodology in sufficient detail (worked examples being usual, but not mandatory) to enable a skilled person to reproduce the invention for himself and show that it works in accordance with the claims of the inventor. Second, the technical description must also describe any new microorganism involved in the invention. Clearly the accession number assigned to the organism by the culture collection in which it has been deposited must be quoted, but beyond that the extent of characterisation required varies between countries. The most extensive requirements are those of the Japanese patent office, which gives in its 'examination standard' a detailed list of the properties which should be recorded. The EPO has less stringent guidelines, and at the other extreme the Netherlands will accept deposit of the organism in lieu of any characterisation. Many countries expect the kind of taxonomic data that would be used in scientific publications, although most do not insist on it and accept deposit as a means of offsetting deficiencies in the written characterisation. In general, an applicant is well advised to provide characterisation data 'to the extent available to him' (Crespi, 1985).

The patent specification – claims. The claims are perhaps the most important part of the patent specification as far as the applicant is concerned, because they set out precisely the extent of the protection being sought. This is particularly so in, for example, the UK and USA where great attention is paid to the exact wording of the claims. Any loophole left here can leave the inventor exposed to competition from which he might otherwise have been protected. In Germany, on the other hand, the claims are viewed less literally in that more attention is given to them as indicators of the basic inventive idea. The EPO adopts a middle course, trying to strike a balance between the rights of the inventor and those of third parties. Again, the reader is referred to Crespi (1982) and to Ruffles (1986) for more detailed discussion about patent claims; only the salient points will be given here.

Normal practice is to make the scope of the initial claims as broad as possible, leaving it to the patent office to object if it believes that too much is being claimed for the invention. In the final event, the applicant may feel that the degree of protection he has been allowed is less than it ought to be, but this is better than finding that he is the author of his own misfortune by having claimed too little in the first place. Nevertheless, the claims must not be extravagant; they must be based on the description and be supported by it. Thus the greater the degree of novelty and ingenuity indicated by the description, and the wider the variety of worked examples given, then the broader the claims that are likely to be accepted.

Claims are usually presented as a numbered set. The first is often the broadest and is the general claim; this is followed by subclaims, each defining by example particular aspects of the general claim, and each generally being narrower in scope than the one before it. The subclaims represent fall-back positions if the broader claims are held invalid. Claims of more than one kind should be included wherever possible, e.g. a new chemical compound, a microbiological process of producing it, the organism used in the process claimed *per se*, a method of diagnosis using the new compound, and a kit incorporating the new compound and conventional reagents for the diagnosis. Table 6.4 gives examples of sets of patent claims relating to different cell types.

Table 6.4. *Examples of sets of patent claims*

US Patent no. 4,567,146

We claim:

1. A recombinant plasmid characterized in that it contains DNA of (1) a first Rhizobium plasmid identifiable as being the same as the plasmid pVW5JI or pVW3JI of lower molecular weight present in the culture of the strain of *Rhizobium leguminosarum* NCIB 11685 or 11683 respectively and (2) a second Rhizobium plasmid found in bacteria of another strain of *Rhizobium*

Table 6.4 (*cont.*)

leguminosarum, said second plasmid having Rhizobium genes coding for nodulation, nitrogen fixation and hydrogen uptake ability but which is non-transmissible.

2. A method of preparing a culture of bacteria of the genus Rhizobium, which method is characterized in that

(1) in a first cross, a donor strain of Rhizobium, containing (a) a Rhizobium plasmid lacking genes coding for nodulation but which is transmissible, is crossed with a recipient strain of Rhizobium, carrying (b) a Rhizobium plasmid having Rhizobium genes coding for nodulation, nitrogen fixation and hydrogen uptake ability but which is non-transmissible, whereby a transconjugant strain carrying a plasmid which is formed from said plasmids (a) and (b) and is a conjugal precursor of a recombinant plasmid (c) having genes coding for nodulation, nitrogen fixation and hydrogen uptake ability and being transmissible is obtained;
(2) said transconjugant strain is separated from donor and recipient strains and cultured to produce a substantially pure culture thereof;
(3) in a second cross, the transconjugant strain from the first cross is used as a donor strain and crossed with a plasmid-containing recipient strain whereby a transconjugant strain carrying a recombinant plasmid (c) is obtained; and
(4) said transconjugant strain from the second cross is separated from donor and recipient strains and cultured to produce a substantially pure culture thereof.

3. A method according to claim 2 characterized in that the transmissible plasmid (a) carries at least one drug-resistance gene.

4. A method according to claim 3 characterized in that the transmissible plasmid is pVW5JI or pVW3JI, identifiable as being the same as the plasmid of lower molecular weight present in the culture of a strain of *Rhizobium leguminosarum* NCIB 11685 (pVW5JI) or NCIB 11683 (pVW3JI), and a kanamycin-resistant transconjugant strain is separated in each cross.

5. A method according to claim 2 characterized in that the transmissible plasmid (a) contains a selectable determinant.

6. A method according to claim 2, characterized in that the donor and recipient strain are of the species *Rhizobium leguminosarum*.

7. A method of impairing hydrogen uptake ability to bacteria of the genus Rhizobium, which method is characterized in that (1) a strain of *Rhizobium leguminosarum* NCIB 11684 or NCIB 11682, as a donor strain, is crossed with a recipient strain of *Rhizobium leguminosarum* to produce a kanamycin-resistant transconjugant strain, said recipient strain being one which permits selection of the transconjugant strain against the donor and recipient strains and which allows the transconjugant strain to be selected against when used as a donor in a subsequent cross with another strain of *Rhizobium leguminosarum*, (2) said transconjugant strain is separated from the donor and recipient strains and cultured to produce a substantially pure culture thereof; (3) in a second cross the transconjugant strain obtained from the first cross is used as a donor strain and crossed with a recipient strain of *Rhizobium leguminosarum* to produce a kanamycin-resistant transconjugant strain and (4) said transconjugant strain from the second cross is separated from the donor and recipient strains to produce a biologically pure culture thereof.

Table 6.4 (*cont.*)

---

8. A method according to claim 7 characterized in that the recipient strain for the first cross is auxotrophic and has resistance to a drug other than kanamycin.

9. A method according to claim 7 or 8 wherein the recipient strain for the second cross is a naturally occurring strain.

10. A Rhizobium plasmid pIJ1008 having Rhizobium genes coding for streptomycin and kanamycin resistance, nodulation, nitrogen fixation and hydrogen uptake properties, which is transmissible and which is the plasmid of lowest molecular weight present in the culture of a strain of *Rhizobium leguminosarum* NCIB 11684 by virtue of the fact that it migrates the fastest on agarose gel in a gel electrophoresis determination in which a gel of 0.7% agarose in Tris-borate buffer of pH. 8.3 is subjected to electrophoresis at 25 mA and 100 volts at 4°C for 16 to 20 hours in the dark.

11. A Rhizobium plasmid pIJ1007 having Rhizobium genes coding for streptomycin and kanamycin resistance, nodulation, nitrogen fixation and hydrogen uptake properties, which is transmissible and which is the plasmid of lowest molecular weight present in the culture of a strain of *Rhizobium leguminosarum* NCIB 11682 by virtue of the fact that it migrates the fastest on agarose gel in a gel electrophoresis determination in which a gel of 0.7% agarose in Tris-borate buffer of pH 8.3 is subjected to electrophoresis at 25 mA and 100 volts at 4°C for 16 to 20 hours in the dark.

12. A biologically pure culture of bacteria of the genus Rhizobium characterized in that it contains a plasmid selected from the group consisting of pIJ1008 and pIJ1007.

13. A culture according to claim 12 of bacteria of the species *Rhizobium leguminosarum*.

14. A biologically pure culture of bacteria of the genus Rhizobium containing a recombinant plasmid characterized in that said plasmid contains DNA of (1) a first Rhizobium plasmid identifiable as being the same as the plasmid pVW5JI or lower molecular weight present in the culture of the strain *Rhizobium leguminosarum* NCIB 11685 or 11683 respectively and (2) a second Rhizobium plasmid found in bacteria of another strain of *Rhizobium leguminosarum*, said second plasmid having Rhizobium genes coding for nodulation, nitrogen fixation and hydrogen uptake ability but which is non-transmissible.

US Patent no. 4,546,082

What is claimed is:

1. A DNA expression vector capable of expressing in yeast cells a product which is secreted from said yeast cells, said vector comprising at least a segment of alpha-factor precursor gene and at least one segment encoding a polypeptide.

2. A DNA expression vector according to claim 1 wherein said segment encoding a polypeptide is an insertion into said alpha-factor precursor gene.

3. A DNA expression vector according to claim 1 wherein said segment encoding a polypeptide is a fusion at a terminus of said alpha-factor precursor gene.

4. A DNA expression vector according to claims 2 or 3 wherein coding sequences for mature alpha-factor are absent from said segment of alpha-factor precursor.

Table 6.4 (*cont.*)

5. A DNA expression vector according to claim 1 wherein said polypeptide is somatostatin.
6. A DNA expression vector according to claim 1 wherein said polypeptide is ACTH.
7. A DNA expression vector according to claim 1 wherein said polypeptide is an enkephalin.
8. A yeast strain transformed with a DNA expression vector of claim 1.
9. A method for producing a DNA expression vector containing alpha-factor gene comprising the steps of
(a) transforming a MAT alpha 2 leu 2 yeast strain with a gene bank constructed in plasmid YEp13;
(b) selecting for leu transformants from the population formed in step (a);
(c) replacing the transformants from step (b) and
(d) screening for alpha-factor producing colonies.
10. A DNA expression vector formed according to the method of claim 9.

UK Patent no. 1,346,051
What we claim is:
1. *Fusarium graminearum* Schwabe deposited with the Commonwealth Mycological Institute and assigned the number I.M.I. 145425 and variants and mutants thereof.
2. *Fusarium graminearum* Schwabe I–7 deposited with the Commonwealth Mycological Institute and assigned the number I.M.I. 154209.
3. *Fusarium graminearum* Schwabe I–8 deposited with the Commonwealth Mycological Institute and assigned the number I.M.I. 154211.
4. *Fusarium graminearum* Schwabe I–9 deposited with the Commonwealth Mycological Institute and assigned the number I.M.I. 154212.
5. *Fusarium graminearum* Schwabe I–15 deposited with the Commonwealth Mycological Institute and assigned the number I.M.I. 154213.
6. *Fusarium graminearum* Schwabe I–16 deposited with the Commonwealth Mycological Institute and assigned the number I.M.I. 154210.
7. Fungal cultures containing a strain of *Fusarium graminearum* Schwabe I.M.I. 145425 or a mutant or variant thereof in a culture medium in which this strain is present in a culture medium containing or being supplied with nutrients or additives necessary for the sustenance and multiplication of the strain, the medium having a pH between 3.5 and 7 and the temperature of the medium being maintained at a precise value within the range of between 25 and 34°C.
8. A method of cultivating a strain of *Fusarium graminearum* Schwabe I.M.I. 145425 or a mutant or variant thereof wherein the strain is present in a culture medium containing or being supplied with nutrients or additives necessary for the sustenance and multiplication of the strain, the medium having a pH between 3.5 and 7 and the temperature of the medium being maintained at a precise value within the range of between 25 and 34°C,
9. A method for the preparation of variants of *Fusarium graminearum* Schwabe I.M.I. 145425 which comprises growing the parent strain I.M.I. 145425 under continuous culture conditions with carbon limitation in a fermentation.
10. A method for the preparation of variants of *Fusarium graminearum* Schwabe I.M.I. 145425 which comprises growing the parent strain I.M.I.

Table 6.4 (*cont.*)

145425 on a glucose based medium at 25° to 30°C under continuous culture conditions at a dilution rate of 0.10 to 0.15 hrs^{-1} with carbon limitation in a fermentation for 1100 hours.

11. A method as claimed in claim 10 wherein the resulting proliferated variants are isolated by dilution plating.

12. A method for the preparation of variants of *Fusarium graminearum* Schwabe I.M.I. 145425 substantially as described with reference to Examples 1 to 5 hereinbefore set forth.

13. Fungal cultures containing *Fusarium graminearum* Schwabe I.M.I. 145425 or mutants or variants thereof substantially as described with reference to any one of Examples 6 to 11 hereinbefore set forth.

UK Patent no. 1,300,391

What we claim is:

1. A human embryo liver cell line having the characteristics of cells deposited with the American Type Culture Collection under number CL99.

2. A cell culture system comprising cells derived from the human embryo liver cell line designated by A.T.C.C. number CL99 in a nutrient culture medium therefore.

3. A virus culture system comprising cells derived from the human embryo liver cell line designated by A.T.C.C. number CL99 inoculated with a virus capable of replication in said cells, and a nutrient culture medium adapted to support growth of the virus-cell system.

4. A culture system according to claim 2 or 3, wherein the nutrient culture medium contains Eagle's minimum essential medium and heat-inactivated foetal calf serum.

5. A culture system according to claim 4, wherein the nutrient culture medium contains Eagle's minimum essential medium, heat-inactivated foetal calf serum, sodium bicarbonate and one or more antibiotics.

6. A virus cultivation process which comprises maintaining a viable culture of cells derived from the human embryo liver cell line designated by A.T.C.C. number CL99 in a nutrient culture medium, inoculating the culture with a virus to which the cells are susceptible and cultivating the virus in the culture.

7. A process according to claim 6, wherein the virus is of the group consisting of adenoviruses, San Carlos viruses, ECHO viruses, arthropodborne group A viruses and other arboviruses, pox viruses, myxoviruses, paramyxoviruses, piconaviruses, herpes viruses and the AR 17 haemovirus.

8. A process according to claim 6, wherein the virus is a hepatitis virus.

9. A process according to claim 7, wherein the virus is of the group consisting of adenovirus types 2, 3, 4, 5, 7 and 17, San Carlos virus types 3, 6, 8 and 49, ECHO virus type 11, Sindbis virus, vaccinia virus, influenza A2 virus and other influenza viruses, Sabin poliovirus type 1 and other poliomyelitis viruses, and the AR–17 haemovirus.

10. A virus whenever cultivated by the process of any of claims 6 to 9.

11. Antigenic material obtained from a virus according to claim 10.

12. A vaccine comprising a virus according to claim 10, in an administrable form and dosage.

13. A vaccine comprising antigenic material according to claim 11, in an administrable form and dosage.

14. A vaccine comprising antibodies produced by a virus according to claim

Table 6.4 (*cont.*)

10 or antigenic material according to claim 11, in an administrable form and dosage.
15. A cell culture system substantially as described in Example 3.
16. A virus cultivation process substantially as described in Example 3.

6.5.3 *Filing the application*

A single application filed in one country will result in patent protection only in that country. Therefore when the disclosure of an invention is likely to lead to serious foreign competition the normal course is to seek protection in several countries. Fortunately this does not have to be done all at once and the first application is usually filed in the applicant's own country. The patent office gives this application a number and, more importantly, a filing date. The significance of this first filing date ('priority date') is that it establishes the priority of the invention; in other words any later applications made in that country by other people for the same invention are pre-empted by it. Furthermore, provided that the applicant files any corresponding foreign applications within 12 months of his basic national filing, the original priority date is also recognised by nearly all overseas countries. However, the same priority date cannot be claimed for material not included in the basic national application (the 'priority document').

The EPO also recognises the original priority date for applications filed with it within 12 months of the basic national filing. The advantage of the European system is that an application filed with the EPO results in a clutch of 'national' patents, valid in those countries party to the EPC that the applicant has designated as being territories in which he wants patent protection. The applicant must, however, designate these countries at the outset; he can drop some of them later by not paying renewal fees, but he cannot add to them. Use of the EPC system is not mandatory in Europe however; if an inventor wishes, he can instead file separate national applications in individual countries. In fact, if protection is not required in more than two or three European countries, the national route may be cheaper.

The first steps along the road to obtaining patent protection involve drafting the patent specification, filing the basic national application and, within 12 months, filing the appropriate foreign applications (redrafting and developing further the original specification if appropriate). As Crespi (1982) has pointed out, a year is not long when account is taken of the need to evaluate the importance of the invention, decide on the extent of foreign patenting desired and implement the decision. Implementation involves drafting the final specification, sending documents around the world and possibly having the specification translated into other languages. In this last respect it is all too easy, say, for the English-speaking applicant to forget that the German Patent Office will expect an application to be written in German. (When filing with the EPO, however, the application may be drafted in English, French or German.) Thus although the patent agent will take care of all the documentary procedures, consulting the applicant where necessary, the latter cannot afford to relax completely at this stage. Attention must also be paid, of course, to ensuring that by the filing date (or, where applicable, the priority date) the microorganism used in the invention has been deposited in a suitable culture collection. This will be discussed in considerable detail later (see Section 6.5.5 below); for the present it is more convenient to follow the progress of the patent application itself.

6.5.4 *Patent office procedures*

Once the final national and/or foreign applications have been filed and the microorganism deposited, the applicant must wait for his application to be processed by the patent office. The exact procedures and the time they take vary widely between countries. Therefore, only a broad outline, illustrated with a few examples, can be given here.

All major patent offices carry out a novelty search, which is usually a patent and literature search, followed by a critical or substantive examination of the application. Under the EPC and many European national systems, the search and examination are treated separately. After the search, a report is sent to the patent agent, pointing out material (including earlier patents) considered relevant to the application. Under the EPC, this report and the patent application itself should be published 18 months after the priority (basic national filing) date, although in practice, delays in the issuance of the search report are common. At this point, the claims can be modified by the applicant if it seems unlikely that they will be accepted as they stand. Also with the publication of the application, the deposited microorganism becomes available (to varying extents) under most European systems (see Release of samples in Section 6.4.2). For the application to proceed further, the applicant must now ask for a substantive examination to be made of it. This request must be made within a specified period or the application will lapse.

Under the Japanese system, patent applications are published after 18 months, but there is no search or examination unless and until the applicant requests it, which he must do within seven years. In Japan and the USA, both the search and substantive examination are carried out before a report is issued to the applicant.

In all systems, a written response to the patent examiner's report must be made within a certain period or the application may fail by default. There usually then follows a variable period of negotiation with the examiner ('prosecution of the application') as to how broad the final claims should be in view of the prior art. If agreement is reached, then the application is accepted. If not, it is refused and the applicant and his agent must then consider whether to pursue an appeal to a higher tribunal.

Negotiations with the patent authorities and the meeting of deadlines are usually taken care of by the patent agent. Unless questions of a highly technical nature are raised, the involvement of the inventor himself in these proceedings is generally minimal.

After an application has been accepted, in most countries the patent is then granted and published (for the first time in the USA and Canada). At this point the deposited microorganism becomes available for the first time in the USA, Canada, Japan and (unless the EPC route has been followed) the Netherlands. Under the EPC system the microorganism, available since the first publication only to an independent expert if the applicant has so opted, becomes generally available.

Major patent offices permit a period immediately after the patent application has been allowed or the patent has been granted for it to be challenged by third parties. The extent of this 'opposition' period varies considerably between countries. Thus, for instance, the EPC allows a nine month period for after-grant revocation of a European patent (before it becomes a collection of national patents). Japan allows an opposition period of three months before grant. In the USA there is no opposition procedure as such. Instead, anyone can ask for a re-examination of any patent, regardless of when it was granted, provided that he can cite pertinent prior art previously unconsidered by the Patent Office and which is sufficient to convince the Office that the issue should be re-opened. Exceptionally, a UK patent can be revoked by application to the Patent Office at any time after grant.

The length of time for which a patent lasts once it has been granted also varies between different countries. In the USA and Canada, for instance, the period is 17 years from the date of grant, regardless of how long the application has been pending the outcome of negotiations between the applicant and the patent office. In Europe, on the other hand, the term is 20 years from the application date. In Japan, it is 15 years from publication for opposition purposes, or 20 years from the application date, whichever is the shorter. Maintenance of the patent for its full term is subject to periodic renewal fees, non-payment of which will result in the patent lapsing.

6.5.5 *Depositing the microorganism*

With an invention involving the use of a new microorganism, that is, one not already available to the public, there is one other vital act to be performed at an early stage of the patenting procedure. The microorganism must be deposited in a suitable culture collection in order to complete the disclosure of the invention. Since in almost all cases the deposit must have been effected at the latest by the filing date (or, where applicable, by the priority date), it might fairly be said that, apart from drafting the specification, this is often the first practical step to be taken towards obtaining the patent. Moreover, it is a step that relies for its effective accomplishment much more on the inventor than on his agent. The latter can do little more than advise about the documentary formalities and deadlines and perhaps suggest appropriate collections. It is the inventor who knows his organism, the technical difficulties in handling it, how long is needed to grow it, and any legal constraints in respect of its pathogenicity, which might delay matters. Thus it is up to the inventor to brief his agent so that between them they can ensure that the culture collection receives the organism in good time to allow for any possible delays or mishaps.

As mentioned above, the mechanism of deposit is now regulated internationally by the Budapest Treaty, and even in countries not yet party to the Treaty, its procedures tend to be viewed as a model system. Nevertheless, for purely national purposes, deposit under the Treaty is often not necessary (see National deposit requirements in Section 6.4.2). However, for the international recognition of a single deposit, using the Budapest Treaty is by far the safest course of action and the following account will be concerned mainly with the Budapest Treaty system. Although the following discussion goes into some detail, much more comprehensive information is contained in the *Guide to the Deposit of Microorganisms for the Purposes of Patent Procedure* issued by the World Intellectual Property Organization (WIPO),

Geneva. For convenience, the term 'depositor' will be used in this connection in preference to 'applicant' or 'inventor'. Lastly, it should be borne in mind throughout that the date of deposit is the date on which the culture collection physically *receives* the culture, rather than the date when the culture is formally accepted.

Requirements of the Budapest Treaty. Under the Budapest Treaty a deposit must be made with an International Depositary Authority (IDA) according to the provisions of Rule 6 of the Treaty. The requirements for making such a deposit are laid down in Rule 6.1(a), which requires that the culture sent to an IDA must be accompanied by a written statement, signed by the depositor and containing the following information:

(i) an indication that the deposit is made under the Treaty and an undertaking not to withdraw it for the period specified in Rule 9.1;

The period specified in Rule 9.1 is five years after the latest request for a sample, and in any case at least 30 years. The important thing to note here is that a deposit made under the Budapest Treaty is permanent and, having made it, the depositor cannot later ask for it to be cancelled, regardless of whether a patent is eventually granted. This applies even if he abandons his patent application.

(ii) the name and address of the depositor;

(iii) details of the conditions necessary for the cultivation of the microorganism, for its storage and for testing its viability and also, where a mixture of microorganisms is deposited, descriptions of the components of the mixture and at least one of the methods permitting the checking of their presence;

This requirement simply ensures that the culture collection is given enough information to enable it to handle the organism correctly. The instructions about (intentionally) mixed cultures are included so that a positive viability statement (see below) is not issued when all the components of the co-culture are not viable.

(iv) an identification reference (number, symbols etc.) given by the depositor to the microorganism;

The term 'identification reference' is sometimes wrongly taken to refer to a taxonomic identification, whereas it simply means 'strain designation'.

(v) an indication of the properties of the microorganism which are or may be dangerous to health or the environment, or an indication that the depositor is not aware of such properties.

The requirements of Rule 6.1(a) are mandatory and cannot be varied either by the depositor or by the IDA. Indeed if the depositor does not comply with them all, the IDA is obliged to ask him to do so before it can accept the deposit. The same does not apply to Rule 6.1(b), which is not really a rule at all but simply an exhortation. According to Rule 6.1(b) 'it is strongly recommended that the written statement . . . should contain the scientific description and/or proposed taxonomic designation of the deposited microorganism'.

As well as the above requirements, the Treaty permits the IDA to set certain conditions of its own (Rule 6.3(a)). These are:

(i) that the microorganism be deposited in the form and quantity necessary for the purpose of the Treaty and these Regulations;

Thus an IDA may require that cultures are submitted to it in a particular state, e.g. freeze-dried, in agar stabs, etc., and that a specified number of replicates is provided.

(ii) that a form established by such authority and duly completed by the depositor for the purposes of the administrative procedures of such authority be furnished;

This refers to the accession form (and any other form) routinely used by the culture collection.

(iii) that the written statement . . . be drafted in the language, or in any of the languages, specified by such authority . . .;

This is an obvious requirement, permitting a Japanese depository, for example, to ask for information to be supplied to it in Japanese.

(iv) that the fee for storage . . . be paid;

(v) that, to the extent permitted by the applicable law, the depositor enter into a contract with such authority defining the liabilities of the depositor and the said authority.

This provides for the IDA to make the kind of contractual arrangements with the depositor that would be usual under the laws of contract of the IDA's own country. Without this provision, some culture collections would have been unwilling to become IDAs.

It is entirely up to the IDA whether it requires any or all of the above from the depositor, but if it does, then the depositor has no option but to comply. Some of the requirements of existing IDAs are summarised in Table 6.5.

Table 6.5. *International Depositary Authorities at 30 November 1989*

International Depositary Authority	Microorganisms accepted	Minimum no. of replicates to be provided by the depositor
SUMMARY		
Agricultural Research Service Culture Collection (NRRL) Peoria USA	Non-pathogenic bacteria, actinomycetes, yeasts, moulds	
American Type Culture Collection (ATCC) Rockville USA	Most kinds	
Australian Government Analytical Laboratory (AGAL) Pymble NSW Australia	Non-pathogenic bacteria, actinomycetes, yeasts and fungi; non-hazardous nucleic acid preparations and phages	
Centraalbureau voor Schimmelcultures (CBS) Baarn Netherlands	Fungi, yeasts, actinomycetes, bacteria	
Collection Nationale de Cultures de Microorganismes (CNCM) Paris France	Bacteria, actinomycetes, fungi, yeasts, viruses, animal and plant cells	

Culture Collection of Algae and Protozoa (CCAP) Ambleside and Oban UK	Algae, non-pathogenic protozoa
Culture Collection of the CAB International Mycological Institute (CMI CC) Kew UK	Non-pathogenic fungi
Deutsche Sammlung von Mikroorganismen (DSM)[b] Braunschweig Federal Republic of Germany	Non-pathogenic bacteria, actinomycetes, fungi, yeasts, phages, plasmids in a host or as a DNA preparation
European Collection of Animal Cell Cultures (ECACC)	Cell lines, animal viruses
Fermentation Research Institute (FRI) Ibaraki-ken, Japan	Non-pathogenic fungi, yeasts, bacteria, actinomycetes, animal and plant cell cultures
Institute of Microorganism Biochemistry and Physiology of the USSR Academy of Science (IBFM) Moscow Region, USSR	Non-pathogenic bacteria including actinomycetes, fungi including yeasts
In Vitro International Inc. (IVI) Linthicum, USA	Most kinds
National Bank for Industrial Microorganisms and Cell Cultures (NBIMCC) Sofia, Bulgaria	Bacteria including actinomycetes, microscopic fungi and yeasts, algae, animal cells, animal viruses, plasmids

Table 6.5. (*cont.*)

International Depositary Authority	Microorganisms accepted	Minimum no. of replicates to be provided by the depositor
National Collection of Agricultural & Industrial Microorganisms (NCAIM) Budapest, Hungary	Non-pathogenic bacteria, fungi, yeasts	
National Collections of Industrial and Marine Bacteria Ltd. (NCIMB) Aberdeen, UK	Non-pathogenic bacteria, actinomycetes, yeasts, phages, plasmids in host cells or as DNA preparations, seeds	
National Collection of Type Cultures (NCTC) London, UK	Pathogenic bacteria	
National Collection of Yeast Cultures (NCYC) Norwich, UK	Non-pathogenic yeasts	
Nationale Sammlung von Mikroorganismen (IMET) Jena, German Democratic Republic	Non-pathogenic bacteria including actinomycetes and cyanobacteria, fungi including yeasts, algae, bacterial viruses, plasmids in hosts or as DNA preparations	
USSR Research Institute for Antibiotics of the USSR Ministry of the Medical and Microbiological Industry (VNIIA) Moscow, USSR	Non-pathogenic bacteria including actinomycetes, fungi including yeasts	

USSR Research Institute for Genetics and Industrial Microorganism Breeding of the USSR Ministry of the Medical and Microbiological Industry
(VNII Genetika)
Moscow, USSR

Non-pathogenic bacteria including actinomycetes, fungi including yeasts

DETAILED INFORMATION

Australia

Australian Government Analytical Laboratories (AGAL)
The New South Wales Regional Laboratory
1 Suakin Street
Pymble, NSW 2073

Bacteria (including actinomycetes), yeasts and fungi with a hazard categorisation no greater than WHO Classification Risk Group 2, that can be preserved without significant change to their properties by the methods of preservation in use (liquid nitrogen storage and lyophilisation).

Nucleic acid preparations and phages may be accepted if the depositor certifies that they pose no hazard when handled by normal laboratory procedures and the depositor supplies suitable material for preservation.

At present, AGAL does not accept for deposit animal, plant, algal and protozoal cultures, cultures of viral, rickettsial and chlamydial agents, microorganisms prohibited by Australian law, or fastidious microorganisms which may require in the view of the curator special attention to handling and preparation for storage.

Bacteria, fungi, yeasts: 6
Phages, plasmids: sufficient quantity and titre for preservation

Table 6.5. (*cont.*)

International Depositary Authority	Microorganisms accepted	Minimum no. of replicates to be provided by the depositor
Bulgaria		
National Bank for Industrial Microorganisms and Cell Cultures (NBIMCC) 125 Lenin Blvd. Block 2 Sofia	Bacteria, actinomycetes, microscopic fungi, yeasts, microscopic algae, animal cell lines, animal viruses and microorganisms containing plasmids.	Bacteria, fungi, yeasts: 3 Viruses, cell lines: 10
France		
Collection Nationale de Cultures de Micro-organismes (CNCM) Institut Pasteur 28 rue du Dr Roux F-75724 Paris Cédex 15	Bacteria (including actinomycetes), bacteria containing plasmids; filamentous fungi and yeasts, and viruses, EXCEPT: – cellular cultures (animal cells, including hybridomas and plant cells): – microorganisms whose manipulation calls for physical insulation standards of P3 or P4 level, according to the information provided by the National Institutes of Health *Guidelines for Research Involving Recombinant DNA Molecules* and *Laboratory Safety Monograph*; – microorganisms liable to require viability testing that the CNCM is technically not able to carry out; – mixtures of undefined and/or unidentifiable microorganisms. The CNCM reserves the possibility of refusing any microorganism for security reasons: specific risks to human beings, animals, plants and the environment.	Cell lines: 12 Other organisms: 8

	In the eventuality of the deposit of cultures that are not or cannot be lyophilised, the CNCM must be consulted, prior to the transmittal of the microorganism, regarding the possibilities and conditions for acceptance of the samples; however, it is advisable to make this prior consultation in all cases.	
Federal Republic of Germany DSM – Deutsche Sammlung von Mikroorganismen und Zellkulturen GmbH Mascheroder Weg 1b D-3300 Braunschweig	Bacteria, including actinomycetes, fungi, including yeasts, bacteriophages, plasmids (a) in a host (b) as an isolated DNA preparation. The following phytopathogenic microorganisms are not accepted for deposit: *Coniothyrium fagacearum; Endothia parasitica; Gloeosporium ampelophagum; Septoria musiva; Synchytrium endobioticum.*	2
German Democratic Republic IMET – Nationale Sammlung von Mikroorganismen IMET-Hinterlegungsstelle Beutenbergstrasse 11 6900 Jena	Strains of bacteria, including actinomycetes and cyanobacteria, fungi, including yeasts, unicellular and filamentous algae, bacterial viruses, plasmids *per se* or included in strains. Strains and materials constituting a danger for man's health or a hazard for the environment, or for the storage or maintenance of which the depository authority is technically not in a position, may be excluded from deposit.	Plasmids: 5 Other organisms: 2

Table 6.5. (*cont.*)

International Depositary Authority	Microorganisms accepted	Minimum no. of replicates to be provided by the depositor
Hungary		
National Collection of Agricultural and Industrial Microorganisms (NCAIM) Department of Microbiology and Biotechnology University of Horticulture and the Food Industry Somlói út 14–16 H-1118 Budapest	Bacteria (including *Streptomyces*) except obligate human pathogenic species (e.g., *Corynebacterium diphtheriae*, *Mycobacterium leprae*, *Yersinia pestis*, etc.). Fungi, including yeasts and moulds, except some pathogens (*Blastomyces*, *Coccidioides*, *Histoplasma*, etc.), as well as certain basidiomycetous and plant pathogenic fungi which cannot be preserved reliably. Apart from the above-mentioned, the following may not, at present, be accepted for deposit: – viruses, phages, rickettsiae; – algae, protozoa; – cell lines, hybridomes.	3 or 25[a]
Japan		
Fermentation Research Institute (FRI) Agency of Industrial Science and Technology Ministry of International Trade and Industry 1-3, Higashi 1-chome Tsukuba-shi Ibaraki-ken 305	Fungi, yeasts, bacteria, actinomycetes, animal cell cultures and plant cell cultures, EXCEPT: – microorganisms having properties which are or may be dangerous to human health or the environment; – microorganisms which require the physical containment level P3 or P4 For experiments, as described in the *Prime Minister's Guidelines for Recombinant DNA Experiments of 1986*.	5

Centraalbureau voor Schimmelcultures (CBS) Oosterstraat 1 Postbus 273 NL-3740 AG Baarn	Fungi, including yeasts, actinomycetes, bacteria other than actinomycetes.	6
United Kingdom		
Commonwealth Agricultural Bureau (CAB), International Mycological Institute (CAB IMI) Ferry Lane Kew, Surrey TW9 3AF	Fungal isolates, other than known human and animal pathogens and yeasts, that can be preserved without significant change to their properties by the methods of preservation in use.	6
Culture Collection of Algae and Protozoa (CCAP) FRESHWATER BIOLOGICAL ASSOCIATION Windermere Laboratory The Ferry House Far Sawrey Ambleside, Cumbria LA22 0LP and SCOTTISH MARINE BIOLOGICAL ASSOCIATION Dunstaffnage Marine Research Laboratory PO Box 3 Oban, Argyll PA34 4AD	(i) Freshwater and terrestrial algae and free-living protozoa (Freshwater Biological Association); and (ii) marine algae, other than large seaweeds (Scottish Marine Biological Association).	6
European Collection of Animal Cell Cultures (ECACC) Vaccine Research and Production Laboratory Public Health Laboratory Service Centre for Applied Microbiology and Research Porton Down Salisbury, Wiltshire SP4 0JG	Cell lines that can be preserved without significant change to or loss of their properties by freezing and long-term storage; viruses capable of assay in tissue culture. A statement on their possible pathogenicity to man and/or animals is required at the time of deposit. Up to and including ACDP Category 3* can be accepted for deposit.	12

* Advisory Committee on Dangerous Pathogens: Categorisation of Pathogens according to Hazard and Categories of Containment, ISBN 0/11/883761/3, HMSO, London.

Table 6.5. (*cont.*)

International Depositary Authority	Microorganisms accepted	Minimum no. of replicates to be provided by the depositor
National Collections of Industrial and Marine Bacteria Ltd (NCIMB) 23 St Machar Drive Aberdeen AB2 1RY	(a) Bacteria including actinomycetes, that can be preserved without significant change to their properties by liquid nitrogen freezing or by freeze-drying (lyophilisation), and which are allocated to a hazard group no higher than Group 2 as defined by the UK Advisory Committee on Dangerous Pathogens (ACDP). (b) Plasmids, including recombinants, either (i) cloned into a bacterial or actinomycete host, or (ii) as naked DNA preparations. As regards (i), above, the hazard category of the host with or without its plasmid must be no higher than ACDP Group 2. As regards (ii), above, the phenotypic markers of the plasmid must be capable of expression in a bacterial or actinomycete host and must be readily detectable. In all cases, the physical containment requirements must not be higher than level II as defined by the UK Advisory Committee on Genetic Manipulation (ACGM) and the properties of the deposited material must not be changed significantly by liquid nitrogen freezing or freeze-drying. (c) Bacteriophages that have a hazard rating and containment requirements no greater than	Bacteria, yeasts, phages: 2 Plasmids (as DNA): 10 ml at 20 mcg/ml Seeds: 2500

those cited in (a) or (b), above, and which can be preserved without significant change to their properties by liquid nitrogen freezing or by lyophilisation.

(d) Yeasts (including those containing plasmids) that can be preserved without significant change to their properties by liquid nitrogen freezing or by freeze-drying, that are allocated to a hazard group no higher than ACDP Group 2, and which require physical containment no higher than level II ACGM.

(c) Seeds that can be dried to a low moisture content and/or stored at low temperatures without excessive impairment of germination potential. The right is reserved to refuse the deposit of seeds where dormancy is exceptionally difficult to break.

The acceptance of seeds by NCIMB and the furnishing of samples thereof are subject at all times to the provisions of the Plant Health (Great Britain) Order 1987, including any future amendments or revisions of that Order.

NCIMB must be notified in advance of all intended deposits of seeds so that it may ensure that all relevant regulations are complied with. Any seeds received without prior notification may be destroyed immediately.

Notwithstanding the foregoing, NCIMB reserves the right to refuse to accept any material for deposit which in the opinion of the Curator presents an unacceptable hazard or is technically too difficult to handle.

Table 6.5. (*cont.*)

International Depositary Authority	Microorganisms accepted	Minimum no. of replicates to be provided by the depositor
	In exceptional circumstances, NCIMB may accept deposits which can only be maintained in active culture, but acceptance of such deposits, and relevant fees, must be decided on an individual basis by prior negotiation with the prospective depositor.	
National Collection of Type Cultures (NCTC) Central Public Health Laboratory 61 Colindale Avenue London NW9 5HT	Bacteria that can be preserved without significant change to their properties by freeze-drying and which are pathogenic to man and/or animals.	1
National Collection of Yeast Cultures (NCYC) AFRC Institute of Food Research Norwich Laboratory Colney Lane Norwich NR4 7UA	Yeasts other than known pathogens that can be preserved without significant change to their properties by freeze-drying or, exceptionally, in active culture.	1
USA		
Agricultural Research Service Culture Collection (NRRL) 1815 North University Street Peoria, Illinois 61604	1. All strains of agriculturally and industrially important bacteria, yeasts, molds and *Actinomycetales*, EXCEPT: (a) *Actinobacillus* (all species); *Actinomyces* (anaerobic/microaerophilic, all species); *Arizona* (all species); *Bacillus anthracis*; *Bartonella* (all species); *Bordetella* (all spe-	1 to 30[a]

cies); *Borrelia* (all species); *Brucella* (all species); *Clostridium botulinum; Clostridium chauvoei; Clostridium haemolyticum; Clostridium histolyticum; Clostridium novyi; Clostridium septicum; Clostridium tetani; Corynebacterium diphtheriae; Corynebacterium equi; Corynebacterium haemolyticum; Corynebacterium pseudotuberculosis; Corynebacterium pyogenes; Corynebacterium renale; Diplococcus* (all species); *Erysipelothrix* (all species); *Escherichia coli* (all enteropathogenic types); *Francisella* (all species); *Haemophilus* (all species); *Herellea* (all species); *Klebsiella* (all species); *Leptospira* (all species); *Listeria* (all species); *Mima* (all species); *Moraxella* (all species); *Mycobacterium avium; Mycobacterium bovis; Mycobacterium tuberculosis; Mycoplasma* (all species); *Neisseria* (all species); *Pasteurella* (all species); *Pseudomonas pseudomallei; Salmonella* (all species); *Shigella* (all species); *Sphaerophorus* (all species); *Streptobacillus* (all species); *Streptococcus* (all pathogenic species); *Treponema* (all species); *Vibrio* (all species); *Yersinia* (all species);

(b) *Blastomyces* (all species); *Coccidioides* (all species); *Cryptococcus neoformans; Cryptococcus uniguttulatus; Histoplasma* (all species); *Paracoccidioides* (all species).

(c) All viral, Rickettsial, and Chiamydial agents.

Table 6.5. (*cont.*)

International Depositary Authority	Microorganisms accepted	Minimum no. of replicates to be provided by the depositor
	(d) Agents which may introduce or disseminate any contagious or infectious disease of animals, humans or poultry and which require a permit for entry and/or distribution within the United States of America. (e) Agents which are classified as plant pests and which require a permit for entry and/or distribution within the United States of America. (f) Mixtures of microorganisms. (g) Fastitious microorganisms which require (in the view of the Curator) more than reasonable attention in handling and preparation of lyophilised material. (h) Phages not inserted in microorganisms. (i) Monoclonal antibodies. (j) All cell lines. (k) Plasmids not inserted in microorganisms. 2. Recombinant strains of microorganisms, strains containing recombinant DNA molecules, strains containing their own naturally occurring plasmid(s), strains containing inserted naturally occurring plasmid(s) from another host, strains containing inserted constructed plasmid(s), and strains containing viruses of any kind, excluding those already	

listed as non-acceptable, only if the deposit document accompanying the microbial preparation(s) includes a clear statement that progeny of the strain(s) can be processed at a Physical Containment Level of P1 or less and Biological Containment requirements meet all other criteria specified by the US Department of Health and Human Services, National Institutes of Health; *Guidelines for Research Involving Recombinant DNA Molecules, December 1978* (*Federal Register*, Vol. 43, No. 247 – Friday, December 22, 1978) and any subsequent revisions.

American Type Culture Collection
(ATCC)
12301 Parklawn Drive
Rockville, Maryland 20852

Algae, animal embryos, animal viruses, bacteria, cell lines, fungi, hybridomas, oncogenes, plant viruses, plasmids, plant tissue cultures, phages, protozoa, seeds, yeasts.

The ATCC must be informed of the physical containment level required for experiments using the host vector system, as described in the 1980 National Institutes of Health *Guidelines for Research Involving Recombinant DNA Molecules* (i.e. P1, P2, P3 or P4 facility). The ATCC, for the time being, will accept only those hosts containing plasmids which can be worked in a P1 or P2 facility.

Certain animal viruses may require viability testing in an animal host, which the ATCC may be unable to provide. In such case, the deposit cannot be accepted. Plant viruses which cannot be mechanically inoculated also cannot be accepted.

Animal viruses, cell lines, naked plasmids: 25
Other organisms: 6
Seeds: 2500

Table 6.5. (*cont.*)

International Depositary Authority	Microorganisms accepted	Minimum no. of replicates to be provided by the depositor
In Vitro International, Inc. (IVI) 611(P) Hammonds Ferry Road Linthicum, Maryland 21090	Algae, bacteria with plasmids, bacteriophages, cell cultures, fungi, protozoa, animal and plant viruses and seeds. Recombinant strains of microorganisms will also be accepted, but IVI must be notified in advance of accepting the deposit of the physical containment level required for the host vector system, as prescribed by the National Institutes of Health *Guidelines*. At present, IVI will accept only hosts containing recombinant plasmids that can be worked in a P1 or P2 facility.	Bacteria, fungi, yeasts: 3 Other organisms: 6 Seeds: 400
USSR		
Institute of Microorganism Biochemistry and Physiology of the USSR Academy of Science (IBFM) Pushchino-na-Oke 142292 Moscow Region	Bacteria (including actinomycetes) and microscopic fungi (including yeasts), also if they are carriers of recombinant DNA, are accepted for deposit, to the exclusion of microorganisms that cause disease in man and animals and microorganisms that have a toxicogenic effect on plants or require them to be quarantined.	5 or 50[a]
USSR Research Institute for Antibiotics of the USSR Ministry of the Medical and Microbiological Industry (VNIAA) Nagatinskaya Street 3-a 113105 Moscow	Bacteria (including actinomycetes) and microscopic fungi (including yeasts) for essentially medical purposes are accepted for deposit, to the exclusion of microorganisms that cause disease in man and animals and microorganisms that are toxicogenic for plants or require them to be quarantined.	5 or 50[a]

USSR Research Institute for Genetics and Industrial Microorganism Breeding of the USSR Ministry of the Medical and Microbiological Industry (VNII Genetika) Dorozhnaya Street No. 8 113545 Moscow	Bacteria (including actinomycetes) and microscopic fungi (including yeasts) for essentially industrial and non-medical purposes are accepted for deposit, to the exclusion of microorganisms that cause disease in man and animals and microorganisms that have a toxicogenic effect on plants or require them to be quarantined.	5	

[a] If depositor's own lyophilised cultures are to be stored and distributed.

These, then, are the official requirements which the depositor must meet. For its part, the IDA also must fulfil certain obligations under the Treaty. In particular it must issue to the depositor an official receipt (the contents of which are laid down by the Treaty) stating that it has received and accepted the deposit and it must, as soon as possible, test the viability of the culture deposited and issue an official statement to the depositor informing him of the result. If the culture proves not to be viable, the deposit is worthless, which can lead to major problems (see below). The IDA must also keep the deposit secret from all except those entitled to receive samples; it must maintain the deposit for the 30 or more years required by the Treaty, checking the viability 'at reasonable intervals' or at any time on the demand of the depositor; it must supply cultures to anyone entitled under the relevant patent law to receive them (provided that the IDA has been given proof of entitlement – see below); it must inform the depositor when and to whom it has released samples; it must be impartial and available to any depositor under the same conditions.

New deposits. If by some mischance a microorganism which was viable when deposited dies during storage, or if indeed for any reason the IDA can no longer supply cultures of it, then the IDA must notify the depositor immediately. The latter then has the option of replacing it (Article 4), and provided he does so within three months, the date on which the original deposit was made still stands. When making a new deposit, the depositor must (under Rule 6.2) provide the IDA with:

(1) a signed statement that he is submitting a culture of the same microorganism as deposited previously;
(2) an indication of the date on which he received notification from the IDA of its inability to supply cultures of the previous deposit;
(3) the reason he is making the new deposit;
(4) a copy of the receipt and the last positive viability statement in respect of the previous deposit;
(5) a copy of the most recent scientific description and/or taxonomic designation submitted to the IDA in respect of the previous deposit;
(6) if the new deposit is being made with a different IDA, all the indications required under Rule 6.1(a) (see above).

With regard to item (6) above, the new deposit can be made with a different IDA if the original IDA is no longer operating as such (either entirely or just in respect of that particular kind of microorganism) or if import/export regulations render the original IDA inappropriate for that particular deposit.

It must be remembered that the provisions for making a new deposit cannot be applied to a microorganism which was shown by the IDA to be non-viable when it was originally deposited. There must have been at least one positive viability statement.

'Converted' deposits. The Budapest Treaty allows (Rule 6.4(d)) for a deposit made outside its provisions to be 'converted' to a Treaty deposit (provided, of course, that the culture collection holding the deposit is an IDA). Where the microorganism was deposited before the culture collection became an IDA the date of deposit of the 'conversion' is held to be the date on which the collection acquired IDA status. Otherwise the date of deposit is the date on which the collection physically received the culture. The procedure for converting a deposit usually involves completing the same forms as are used for making a deposit *de novo*. However, only the original depositor (or his successor) can convert a deposit. In all other cases, a separate deposit of the same organism must be made under the Treaty.

Conversion is a useful facility because it means that an earlier non-Budapest Treaty deposit can be accorded the international recognition which it might not otherwise command. Conversion is essential for the recognition by the Japanese patent office of any non-Budapest Treaty deposit made outside Japan.

Guidelines for deposit. It is generally recognised that many depositors (and sometimes their patent agents) are unlikely to be familiar with the minute details of the Budapest Treaty and may not be aware of their obligations in respect of it. Therefore, the forms which IDAs ask prospective depositors to fill in are generally so designed that by completing them correctly the depositor automatically provides all the information required of him by the Treaty. These forms vary to some extent between IDAs, but they all follow a similar general pattern. Any IDA will supply specimens of its forms on request.

Making a deposit under the Budapest Treaty should be quite straightforward, but problems can and do arise. It has to be said that many of these are of the depositor's (or his agent's) own making and that they can be avoided by adhering to a few simple guidelines. Perhaps the most important thing to be remembered is that the

Budapest Treaty procedures take a certain amount of time to complete, even when they are operating ideally. Thus although in principle a deposit does not have to reach the IDA until the filing (or priority) date of the relevant patent application, in practice the wise depositor will start the depositing procedure in good time to allow for any possible delays. Thus if he is intending to deposit in a foreign IDA, say, he should bear in mind any import or quarantine regulations. For instance, it can take several weeks, or even months, to obtain a permit to import cell lines and viruses into the USA. Last-minute deposits are unwise for several reasons, some of the most common being:

(1) postal delays: the culture fails to arrive in time;
(2) customs delays: with deposits from overseas, depositors have not provided adequate shipping information;
(3) the deposit is not the kind of microorganism accepted by the IDA (see Table 6.5);
(4) the microorganism cannot be recovered from the package, e.g. because the culture tube is broken;
(5) the deposit proves to be non-viable; if a microorganism is found by the IDA from the outset to be non-viable, the original date of deposit cannot be applied to any replacement (see above).

For most bacteria, fungi, yeasts, algae and protozoa, viability testing usually takes 3–5 days; for animal cell lines a week or slightly longer is normal; and for animal viruses and plant tissue cells, up to a month is not unusual.

It cannot be emphasised too strongly that however good the depositor's intentions may be, patent offices recognise only the actuality of the deposit. With all this in mind the prudent depositor will also pay attention to a few more elementary points to ensure a timely and trouble free deposit. He will ensure that the microorganism he wishes to deposit is one of the kinds that the IDA he had chosen can officially accept under the Budapest Treaty (see Table 6.5). If there are likely to be technical problems with the organism he will advise the IDA in advance. He will check the administrative and technical requirements of the chosen IDA and ask for the appropriate patent deposit forms, which he will then fill in completely and correctly, for by doing so he should automatically comply with the requirements of Rule 6.1(a) (see above). Although Rule 6.1(a) states that the microorganism should be *accompanied* by a written statement (the completed deposit form), in practice it is often helpful to an IDA to receive the written information in advance of the microorganism itself, so that arrangements can be made to deal with the deposit promptly. This is particularly helpful if, say, a special growth medium has to be prepared by the IDA. Lastly, if the depositor's patent agent is likely to be communicating with the IDA, the depositor should let the IDA know, otherwise it may withhold information until it has ascertained the agent's right to receive it.

Depending on its policy and on the kind of material being deposited, an IDA may or may not prepare subcultures for eventual distribution. Thus in the case of cell lines and naked plasmids (not cloned into a host), for instance, the depositor is usually required to supply sufficient material for the IDA to distribute direct. On the other hand, for bacteria, yeasts, moulds, etc. (with or without plasmids) it is more usual for the IDA to distribute its own preparations. In this case, many IDAs will ask the depositor to check the authenticity of their preparations – a fairly normal culture collection practice. The depositor is not *obliged* by the Budapest Treaty to check these preparations, but he is well advised to do so to ensure that the cultures to be sent out by the IDA will in fact do what is claimed for them in the patent application.

The official aspects of the depositing procedure end with the issuing of the receipt and viability statement by the IDA. These are important for they are the documentary proof that on a particular date a viable deposit has been made according to the terms of the Budapest Treaty.

Technically the receipt should be issued first, but in practice many IDAs find it more convenient to await the results of the viability test and then send out the receipt and viability statement together. In general, for deposits of most bacteria, fungi, yeasts, algae and protozoa, the depositor could expect an IDA to send him both documents within a few days of it having received the deposit. For animal cell lines a week or slightly longer would be normal, and for animal viruses and plant tissue cells four or five weeks would be more usual.

6.5.6 *Obtaining a sample of a patent deposit*

So far in this account the deposit procedure has been considered primarily from the viewpoint of the depositor. It would be useful now to look briefly at the procedures which a third party must follow in order to obtain a culture, since the whole point of deposit is to make the microorganism available.

It is generally admitted that culture collections can neither be expected to be familiar with the patent laws of countries throughout the world nor to know what stage patent applications relating to the deposits they hold have reached. Thus to require a collection to judge for itself whether a particular person is legally entitled to a culture of a particular deposit is considered by many to impose an unfair burden on the collection. Therefore the Budapest Treaty attempts to place the onus on patent offices to ensure that IDAs are not put in this position (Rule 11.3). Patent offices in countries whose laws require that deposited microorganisms must be available without restriction to anyone once the relevant patents have been granted and published can notify the IDAs from time to time of the accession numbers of the strains cited in these patents. However, this provision is not usually adopted. The US Patent Office, for instance, directs that a microorganism must be available from the date of issuance of the relevant US patent, but it does not advise the IDA of this date. Since the issuance or not of a US patent is a simple matter of fact, in the event of any request an IDA merely has to ascertain this fact, either from the requesting party or the depositor. In the author's experience this does not cause any major problems, although the actual furnishing of the sample may be delayed slightly while evidence of publication is obtained. The IDA can then meet any requests for the strains in question without the need for evidence of entitlement.

In cases where the availability of the microorganism is restricted and/or where evidence of entitlement to receive a sample is required, anyone requiring a culture must either obtain the written authorisation of the depositor or he must obtain from a relevant patent office a certificate stating:

(1) that a patent application in respect of the strain in question has actually been filed with that office;
(2) whether the application has been published;
(3) that the person requesting the culture is legally entitled to receive it and has met any conditions that the law requires.

On receipt of a request accompanied by such a certificate, or by the written authorisation of the depositor, the IDA will supply the culture (subject to its normal fee for such cultures being paid). At the same time, the IDA will inform the depositor when and to whom it has supplied the culture, as it is obliged to do by Rule 11.4(g) of the Treaty, unless the depositor has specifically waived this right to be informed.

Except where the direct authorisation of the depositor has been sought, the request for a culture must be made on an official form which can be had from the patent office(s) with which the relevant application has been filed. Most IDAs also have copies of these forms. Thus, the procedure for obtaining a culture of a microorganism deposited under the Budapest Treaty is:

(1) ask the appropriate patent office, or the IDA, for a copy of the form to be used for requesting samples of microorganisms deposited under the Budapest Treaty;
(2) complete the part of the form to be filled in by 'the requesting party';
(3) send the entire form to the patent office, *not* to the IDA;
(4) when the form bearing the appropriate stamp of authorisation is received back from the patent office, send it to the IDA along with a normal purchase order.

Procedures for obtaining organisms deposited for patent purposes outside the Budapest Treaty vary according to the national law. In such cases, the culture collection will have been informed by the depositor or the patent office of the appropriate requirements and should be able to advise accordingly.

It must be remembered that the procedures outlined above relate only to the right to receive cultures according to patent law. They do not over-ride any requirements to be met in respect of import and quarantine regulations, health and safety procedures, plant disease regulations, etc. Thus as well as obtaining patent office authorisation, a person requesting a culture must also ensure that he has obtained any permit or licence necessary for handling the organism in question.

6.6 Further reading

Adler, R. G. (1984). Biotechnology as an intellectual property. *Science* **224**, 357–63.

Anonymous (1982). Japanese Patent Office guidelines for examination of inventions of microorganisms. *Yuasa and Hara Journal* **9** (3).

Beier, F. K., Crespi, R. S. & Straus, J. (1985). *Biotechnology and Patent Protection: An International Review*. Paris: Organization for Economic Co-operation and Development.

Biggart, W. A. (1981). Patentability in the United States of microorganisms, processes utilizing microorganisms, products produced by microorganisms and microorganism mutational and genetic information techniques. *IDEA, Journal of Law and Technology* **22**, 113–36.

Budapest Treaty (1981). *Budapest Treaty on the International Recognition of the Deposit of Microorganisms for the Purposes of Patent Procedure* 1977 and *Regulations* 1981. Geneva: World Intellectual Property Organization.

Byrne, N. J. (1979). Patents on life. *European Intellectual Property Review* **1**, 279–300.

Byrne, J. J. (1983). The agritechnical criteria in plant breeders' rights law. *Industrial Property* (1983), 294–303.

Convention (1980). *Convention of the Grant of European Patents.* (European Patent Convention) 1973 with 1980 amendments. Munich: European Patent Organization.

Cooper, I. P. (1982). *Biotechnology and the Law.* New York: Clark Boarman Co.

Crespi, R. S. (1981). Biotechnology and patents – past and future. *European Intellectual Property Review* **3**, 134–40.

Crespi, R. S. (1982) *Patenting in the Biological Sciences.* Chichester: John Wiley & Sons.

Crespi, R. S. (1985a) Biotechnology patents – a case of special pleading? *European Intellectual Property Review* **7**, 190–3.

Crespi, R.S. (1985b) Microbiological inventions and the patent law – the international dimension. *Biotechnology and Genetic Engineering Reviews* **3**, 1–37.

Crespi, R. S. (1985c) Patent protection in biotechnology: questions, answers and observations. In *Biotechnology and Patent Protection,* ed. F. K. Beier, R. S. Crespi & J. Straus, pp. 36–85. Paris: Organization for Economic Co-operation and Development.

Crespi, R. S. (1986). Patent issues in biotechnology. In *Biotechnology and Crop Improvement and Protection,* British Crop Protection Council Monograph No. 34, ed. Peter R. Day, pp. 209–17.

In re Diamond & Chakrabarty (1980). *US Patents Quarterly* **206**, 193.

Duffy, J. I. (1980). *Chemicals by Enzymatic and Microbial Processes. Recent Advances.* Park Ridge, New Jersey: Noyes Data Corp.

Halluin, A. P. (1982). Patenting the results of genetic engineering research: an overview. In *Patenting of Life Forms,* Banbury Report No. 10, pp. 67–126. Cold Spring Harbor, New York: Cold Spring Harbor Laboratory.

Hüni, A. (1977). The disclosure in patent applications for microbiological inventions. *International Review of Industrial Property and Copyright Law* **8**, 500–21.

Hüni, A. & Buss, V. (1982). Patent protection in the field of genetic engineering. *Industrial Property* (1982), 356–68.

International Convention (1978). *International Convention for the Protection of New Varieties of Plants* 1961, revised 1972. Geneva: World Intellectual Property Organization.

Irons, E.S. & Sears, M. H. (1975). Patents in relation to microbiology. *Annual Review of Microbiology* **29**, 319–32.

In re Lundak (1985). *US Patents Quarterly* **227**, 90.

Plant, D. W., Reimers, N. J. & Zinder, N. D. (eds.) (1982). *Patenting of Life Forms.* Banbury Report No. 10. Cold Spring Harbor, New York: Cold Spring Harbor Laboratory.

Pridham, T. G. & Hesseltine, C. W. (1975). Culture collections and patent depositions. *Advances in Applied Microbiology* **19**, 1–23.

Straus, J. (1985). *Industrial Property Protection of Biotechnological Inventions: Analysis of Certain Basic Issues.* Document BIG/281. Geneva: World Intellectual Property Organization.

Teschemacher, R. (1982). Patentability of microorganisms *per se*. *International Review of Industrial Property and Copyright Law* **13**, 27–41.
Wegner, H. C. (1979). Patenting the products of genetic engineering. *Biotechnology Letters* **1**, 145–40 and 193.
Wegner, H. C. (1980). The Chakrabarty decision patenting products of genetic engineering. *European Intellectual Property Review* **2**, 304–7.
WIPO (1980) *Records of the Budapest Diplomatic Conference for the Conclusion of a Treaty on the International Recognition of the Deposit of Microorganisms for the Purposes of Patent Procedure*. WIPO Publication no. 332 (E), p. 119. Geneva: World Intellectual Property Organization.
WIPO (1986). *Report adopted by the 2nd session of the Paris Union Committee of Experts on Biotechnological Inventions and Industrial Property*. WIPO Document BioT/CE/II/3. Geneva: World Intellectual Property Organization.
WIPO (1988). *Guide to the Deposit of Microorganisms for the Purposes of Patent Procedure*.

The author gratefully acknowledges the helpful comments and criticisms made by many colleagues during the writing of this chapter. Particular thanks go to Mrs B. A. Brandon of the American Type Culture Collection and Mr R. S. Crespi of the British Technology Group. A singular debt of gratitude is owed to Mr R. K. Percy of the British Technology Group, for his extensive advice and painstaking correction of the original draft.

7

Culture collection services

D. ALLSOPP and F. P. SIMIONE

7.1 Introduction

In response to the needs of users, many culture collections provide a range of services to the scientific, technological and commercial world. This chapter provides an introduction to the types of services available from culture collections, but it is beyond its scope to give a comprehensive list of such services. As the range of work that can be undertaken is increasing at many of the collections, the reader should contact individual collections to find out whether they can offer particular services.

7.2 Types of services

7.2.1 *Directly associated and customer services*

The two major services which are intrinsically part of culture collection work are those concerning the identification and preservation of organisms. Collections of necessity need expertise in these fields to be able to function, and many provide comprehensive services in these areas. Aspects of culture identification methods (Chapter 5), sales of cultures (Chapter 3), preservation techniques (Chapter 4) and patent deposits (Chapter 6) are covered elsewhere in this volume.

Safe-Deposits. Many collections hold organisms which are not listed in their catalogues. These cultures are held for a variety of reasons: the organisms may not be fully identified, their taxonomic status may be unclear, their stability in preservation may be suspect or they may be held at the request of the depositor who wishes to have back-up material and yet retain ownership and confidentiality, not releasing the strain to other parties. Many collections have introduced safe-deposit services as a back-up to the depositor's working collection, providing a service intermediate between an open collection deposit and a deposit for patent purposes. Such services enable the depositor to have important organisms professionally preserved and maintained even if the collection would not normally be interested in accessioning them. There is obvious merit in the safekeeping of cultures while they are the subject of research, especially as many laboratories do not have optimum preservation facilities. Most collections make a charge for safe-deposits to cover the long-term storage costs and quality control procedures that are required.

Advice on strain selection. Collections are able to give advice on the selection of strains for special purposes. Such services may involve collection staff in a considerable amount of work and are limited by the sources of information available. In the past such services have relied upon the individual expertise of collection staff and their personal knowledge of the scientific literature. With the development of computer databases and strain data networks (Chapter 2), such services are becoming more frequent and are increasing in efficiency. However, as databases are searched electronically and appropriate microorganisms selected, the expertise of the collection staff is still needed to draw attention to closely allied genera and species that might merit study or to point out the idiosyncracies of individual strains. These advisory services are being placed on a more formal basis and charges may be made.

Advice on maintenance of organisms. Most collections are able to provide information on preservation systems, either on a formal or informal basis. Some of the major collections have produced substantial publications covering the preservation of their own groups of organisms and these may be consulted. However, if any doubt exists or if different media are being tried, the collections can always be consulted for advice. Advisory sheets on particular preservation techniques, or related topics such as the handling of pathogens, elimination of contamination or mite infestation, are often available on request.

Biochemical services. Some groups of organisms, such as bacteria and yeasts, are identified using biochemical tests, and such methods are increasingly used for other organisms, particularly the filamentous fungi. The need for taxonomic clarification using biochemical criteria goes hand in hand with an increasing requirement to provide metabolic and other physiological data to users of the collections. Applied biological industries, and in particular biotechnology, increasingly seek organisms on the basis of activity rather than name, and collections are responding to this requirement.

Collections now enhance their catalogues with strain data, and supply information to strain databases and computer networks as a routine procedure. Some collections also carry out custom-designed screening programmes to select strains with specified attributes for individual clients, particularly in the fields of enzyme and secondary metabolite (including toxin) production or specific growth requirements.

These biochemical services serve both as directly associated and contract services of culture collections (see also Research and development work in section below).

7.2.2 *Contract services*

The services outlined in Section 7.2.1 above usually exist as a consequence of the collection's normal activities; however, in recent years there has been an expansion in other services offered by collections for a variety of reasons. For example, many collections are associated with other institutions, such as research organisations, taxonomic institutes, educational institutions, or commercial firms, any of which may make specific demands on the services of the collection. Again, the collection may be supported by funds provided from external sources which may require particular expertise or services to be developed. Many collections are currently under financial pressure to increase their earning potential, and this has stimulated the development of income-generating activities.

Biological testing. Many modern standards and specifications require the use of microorganisms and cell lines in the testing of products. These include the mould growth resistance of materials (Fig. 7.1), the testing of disinfectants against a range of bacteria, the assessment of mutagenicity of materials against a range of organisms, and toxicity testing. For economic and other reasons there is strong pressure to move away from the use of live animals in the testing of products and towards the use of microorganisms or cell lines instead.

These tests can be carried out in any suitably equipped laboratory, but culture collections are in a very favourable position to carry out such testing themselves, using the organisms or cell lines which they would normally supply to outsiders for such work; indeed, collections may be identified in the standards themselves as an approved source

Fig. 7.1. Chamber used for the testing of industrial products for resistance to mould growth at the CAB International Mycological Institute (IMI).

of such material. Unless a company is equipped for routine testing as part of its quality control procedures, it is often more cost effective for such work to be carried out in a specialised laboratory. To establish a laboratory and train staff to carry out extensive biological testing infrequently would be very expensive, and not necessarily satisfactory from a technical point of view. However, an economic service can be offered by culture collections equipped to carry out such work on a regular basis.

Examples of testing standards using microorganisms and cell lines include British (BS), European (ISO) and USA (ASTM, USP) standards for fungal resistance testing, sterility testing, preservative effectiveness, toxicity and biocompatibility testing (Table 7.1). Culture collections provide reference cultures used in clinical laboratory standards procedures (e.g. NCCLS, ECCLS), and can provide uniform, quality assured sets of reference cultures for use in on-site biological testing for both clinical and industrial applications.

Table 7.1. *Examples of testing standards involving microorganisms, cell cultures and related materials*

British Standards	
BS 1982	Methods of testing for fungal resistance of manufactured building materials made of, or containing, materials of organic origin
BS 2011	The environmental testing of electronic components and electronic equipment. Test J. Mould growth
BS 28458	Flexible insulating sleeving for electrical purposes. Section 12. Mould growth
BS 3046	Specification for adhesives for hanging flexible wall coverings. Appendix G. Test for susceptibility to mould growth
BS 4249	Specification for paper jointing. Section 5.7. Resistance to mould growth
BS 5980	Specification for adhesives for use with ceramic tiles and mosaics. Section 7. Resistance to mould growth
BS 6009	Wood preservatives. Determination of toxic values against wood destroying *Basidiomycetes* on an agar medium
BS 6085	Methods of testing for the determination of the resistance of textiles to microbiological deterioration
European Standards	
ISO 846	Plastics – determination of behaviour under the action of fungi and bacteria – evaluation or measurement of change in mass or physical properties

See also NFX41–514 and DIN 53739 for similar methods of plastics testing in France and Germany, respectively.

Table 7.1 (*cont.*)

US Standards	
ASTM D2574	Standard test method for resistance of emulsion paints in the container to attack by microorganisms
ASTM D3273	Resistance to growth of mould on the surface of interior coatings in an environmental chamber. Standard test method for evaluating the degree of surface disfigurement of paint films by fungal growth or soil and dirt contamination
ASTM G21–70	Standard recommended practice for determining resistance of synthetic polymeric materials to fungi
ASTM G22–76	Standard recommended practice for determining resistance of plastics to bacteria
FDA 21CFR610.12	Sterility testing of biological products
FDA 21CFR610.30	Detection of mycoplasma contamination
MIL STD 810D	Environmental test methods. Method 508.2 Fungus
NCCLS M2–A3	Performance standards for antimicrobial disk susceptibility tests
NCCLS M7A	Methods for dilution antimicrobial susceptibility tests for bacteria that grow aerobically
NCCLS M11A	Reference agar dilution procedure for antimicrobial susceptibility testing of anaerobic bacteria
USDA 9CFR113.26	Detection of viable bacteria and fungi in biological products
USDA 9CFR113.28	Detection of Mycoplasma contamination
USDA 9CFR113.29	Determination of moisture content in desiccated biological products
USP	Antimicrobial preservatives – effectiveness assay
USP	Microbial limits tests
USP	Sterility tests

Consultancy. Many collections have staff with expertise in specialised areas commensurate with their collection responsibilities, who can be made available for consultancy work. Such work is often a forerunner of detailed investigations, or a research programme, and can be time consuming. Since collection staff time for this work is limited, consultancy work is usually offered on a fully charged basis.

Research and development work. Culture collections equipped for testing work and consultancy work may also be involved in research and development work and in industrial investigations which require laboratory facilities. It is often difficult for collections to specify exactly what kind of work they would be prepared to accept, as many problems are unique and may be carried out on a one-off basis. It is usual, therefore, for collections to consider any type of work which falls in their general area of competence. By their very nature, culture collections usually have a wide range of scientific and industrial contacts, and even if they are not able to carry out particular investigations themselves, they may be able to advise on places where such work could be carried out. Culture collections can therefore act as referral centres and this function should not be overlooked.

In addition to research topics on taxonomy and the preservation of organisms, major culture collections are often well placed to undertake other research, often allied to biotechnology, which may be initiated by short-term contracts and testing work. Potential topics for such longer-term research and development work might include:

(1) screening large numbers of isolates for particular biochemical properties and end uses;
(2) comparative studies in regard to enzyme or metabolite production of different strains of the same or closely related species;
(3) development and evaluation of rapid detection procedures for compounds produced by organisms, including the development of commercial kits;
(4) studies on the use of microorganisms as bio-control agents against insect pests, weeds and deteriogenic microorganisms;
(5) selection of test organisms used in the evaluation of materials;
(6) studies on growth requirements and the testing of bioreactors;
(7) evaluation of media, culture vessels, diagnostic reagents and procedures.

Custom preparations. Culture collections may be requested to provide bulk amounts of their organisms on an *ad hoc* basis to commercial organisations with limited facilities. Often this is an efficient and cost-effective method of obtaining bulk inocula, since the collection has the expertise in growing the organisms and ready access to the media.

Government agencies may request multiple units of mixtures of cultures for laboratory certification and proficiency testing in clinical laboratories. Manufacturers of diagnostic instruments may specify cultures with known properties for calibration of their instruments.

Resource development. In the United States the Federal Government has used culture collections extensively to develop research reagents/

cultures. These have been developed through joint efforts with the research scientists defining the needs and quality controls required and the banking and distribution aspects handled by the collections.

7.2.3 *Confidentiality*

Work performed by culture collections can be carried out on a confidential basis if required. Formal mechanisms for ensuring confidentiality are in fact required by some laboratory accreditation schemes (see below) and are already in place in a number of collections. In some areas, such as identification services, users are sometimes content for material sent in to become part of the open collection, should the collection wish to retain it. However, in collections which deal with organisms of industrial importance, it may be more normal to carry out the work in confidence, with all material being destroyed after examination. Enquiries can be treated as confidential, and collections should be questioned about their policy and procedures regarding confidentiality. Submissions are a source of new organisms for building up the resources available from the collections, and some collections will identify cultures for a lower fee if the culture can be retained in the collection.

7.2.4 *Laboratory accreditation*

In many countries national schemes exist for the accreditation of testing laboratories. Such schemes aim to establish standards for the accuracy and efficiency of measuring instruments, quality control procedures, record-keeping, and administration of the laboratory. Major companies may have their own accreditation schemes for laboratories that they use, but where national schemes exist, companies often accept the accreditation of the scheme and do not carry out individual laboratory accreditation on their own behalf. National schemes may also be recognised on an international basis and thus ease the way for laboratories to be accepted on a much wider geographical basis. There are now a sufficient number of national and international testing standards involving microorganisms to make such accreditation schemes worthwhile for culture collection laboratories to adopt.

7.3 Workshops and training

Many culture collection staff are involved in educational activities, either because of their general scientific background or their detailed knowledge of culture collection practices. Instruction may be external, where staff are involved in training courses at local colleges, polytechnics or research institutes in their own country or overseas. Increasingly, however, collections are devising formal programmes of in-house courses. These may be arranged in direct response to a need identified either by the collection itself or by an outside body. In some parts of the world the educational system is such that credits towards a qualification can be gained by attendance at approved outside courses, and where such a system exists, it can ease the task of the culture collection in operating such courses.

Courses vary in length from several weeks' or even months' duration (if dealing with topics such as identification), to one-day lecture courses or seminars on specific industrial or commercial topics. Outside speakers and instructors may be used to enable topics to be covered which extend beyond the expertise available in the collection itself. In addition to formal programmed and advertised courses, it is often possible to obtain individual training within culture collections to suit a particular need. Some collections have special facilities and staff devoted to training programmes, while others may use external facilities at educational institutions nearby. Charges are usually made for training, but in some cases the costs may be subsidised. Advice may be available from collections on suitable sources of funding for prospective students and trainees, especially from developing countries. Some culture collections are either owned by academic institutions such as universities, or are officially associated with them, enabling them to offer training towards MSc and PhD degrees by research.

Some of the training offered is directly related to the normal activities of the collection, and instruction in preservation and maintenance techniques is often available; indeed, such training is often difficult to obtain from other sources. The training facilities offered by culture collections are often much greater than is generally known and the World Federation for Culture Collections' Education Committee is developing a list of teachers, their special expertise and courses available throughout the world (see Chapter 8). Individual collections may also provide information on training facilities in their scientific speciality or geographical area.

7.4 Publications, catalogues and publicity material

The essential publication of any culture collection is its catalogue. Traditionally, these have been produced as hard-copy items but are now becoming available in a computerised on-line form (see Chapter 2). It is always worthwhile contacting a collection if an organism does not appear in its most recent hard-copy catalogue, as additional material may well be available, or information can be given on suitable organisms in reserve collections, which could possibly be released. In addition to catalogues, a few of the major collections produce scientific publications of their own (guides on preservation and maintenance, safety and handling, industrial uses and teaching) as well as articles in the scientific press and monographic studies. Details of such publications may be found in the collections' brochures or newsletters, or in bibliographic databases.

Collection brochures are generally available free of charge, and it is worthwhile asking to be included on the collection's mailing list to ensure receipt of up-to-date information. Despite the fact that collections grow and change with time, enquirers or customers often rely on data from back issues of catalogues which may be many years old. This can lead to considerable confusion and users should ensure that they have the current catalogues before ordering cultures. These problems are minimised as catalogue and strain data become widely available through computer networks. For many people, however, printed information remains the most important reference material.

7.5 Fees and charges

The ways in which culture collections are funded are extremely diverse. Very few culture collections exist as straightforward commercial entities; they are almost all subsidised in some fashion, either directly or indirectly, and the charges made for cultures do not reflect the true cost of production. Nevertheless, many collections offer discounts for bulk orders, special sets, or regularly ordered organisms such as those used for testing and teaching.

The charges made for services, however, more accurately reflect the true cost, though they usually represent good value in comparison with totally commercial services, particularly in the areas of training. Charges for consultancy work, for testing and for laboratory services are normally at competitive commercial rates.

Over the last few years, several important stimuli have been applied to culture collections, including the advent of biotechnology, the development of computerised databases and a harsher economic climate. These and other factors have led culture collections to examine the services they provide and to develop them to cater to the changing demand of their users. Expansion and diversification of the range of services has resulted.

7.6 Suggested reading

Alexander, M., Daggett, P.-M., Gherna, R., Jong, S., Simione, F. & Hatt, H. (1980). *American Type Culture Collection Methods. Laboratory Manual on Preservation: Freezing and Freeze Drying*. Rockville, Maryland: American Type Culture Collection.

Allsopp, D. (1985). Fungal culture collections for the biotechnology industry. *Industrial Biotechnology* **5**, 2.

Allsopp, D. & Seal, J. J. (1986). *Introduction to Biodeterioration*, 136 pp. London: Edward Arnold.

Batra, L. R. & Iijima, T. (eds.) (1984). *Critical Problems for Culture Collections*, 71 pp. Osaka, Japan: Institute for Fermentation.

Cour, I. G., Maxwell, G. & Hay, R. (1979). Tests for bacterial and fungal contaminants in cell cultures as applied at the ATCC. *TCA Manual* **5**, 1157–60.

Dilworth, S., Hay, R. & Daggett, P.-M. (1979). Procedures in use at the ATCC for detection of protozoan contaminants in cultured cells. *TCA Manual* **5**, 1107–10.

Hawksworth, D. L. (1985). Fungus culture collections as a biotechnological resource. *Biotechnology and Genetic Engineering Reviews* **3**, 417–53.

Hay, R. J. (1983). Availability and standardization of cell lines at the American Type Culture Collection: Current status and prospects for the future. In *Cell Culture Test Methods*, STP 810, ed. S. A. Brown, pp. 114–26. Philadelphia: American Society for Testing and Materials.

Jewell, J. E., Workman, R. & Zelenick, L. D. (1976). Moisture analysis of lyophilized allergenic extracts. In *International Symposium on Freeze-Drying of Biological Products. Developments in Biological Standardization* **36**, 181–9.

Kelley, J. (1985). The testing of plastics for resistance to microorganisms. In *Biodeterioration and Biodegradation of Plastics and Polymers* ed. K. J. Seal, pp. 111–24. Cranfield, UK: Cranfield Press.

Kelley, J. & Allsopp, D. (1987). Mould growth testing of materials, components and equipment to national and international standards. *Society of Applied Bacteriology, Technical Series* 23. Oxford, UK: Blackwell Scientific Publications.

Lavappa, K. D. (1978). Trypsin-Giemsa banding procedure for chromosome preparations from cultured mammalian cells. *TCA Manual* **4**, 761–4.

Macy, J. (1978). Identification of cell line species by isoenzyme analysis. *TCA Manual* **4**, 833–6.

Macy, J. (1979). Tests for mycoplasmal contamination of cultured cells as applied at the ATCC. *TCA Manual* **5**, 1151–5.

May, M. C., Grim, E., Wheller, R. M. & West, J. (1982). Determination of residual moisture in freeze-dried viral vaccines: Karl Fischer, gravimetric thermogravimetric methodologies. *J. Biological Standardization* **10**, 249–59.

8

Organisation of resource centres

B. E. KIRSOP and E. J. DaSILVA

8.1 Introduction

Individual resource and information centres provide valuable services to biotechnology, but their role can be substantially enhanced if their activities are effectively co-ordinated. This has been recognised in the past, and a number of committees, federations and networks have been set up for this purpose at the national, regional and international levels. Although the origins and composition of existing organisations differ and their geographical locations are widespread, their common purpose is to support and develop the activities of resource and information centres for the benefit of microbiology.

8.2 International organisation

8.2.1 *World Federation for Culture Collections*

There are fewer difficulties in setting up national and regional co-ordinating mechanisms than international systems, and yet one of the first developments in this area was the formation of the World Federation for Culture Collections (WFCC). In 1962 at a Conference on Culture Collections held in Canada it was recommended that the International Association of Microbiological Societies (IAMS) set up a Section on Culture Collections. The Section was established in 1963. Five years later, at an International Conference on Culture Collections in Tokyo, the formation of the WFCC was proposed and an *ad hoc* committee, together with the Section on Culture Collections, drew up statutes which were agreed at a congress in 1970. Following the conversion of the IAMS to Union status, the WFCC is now a federation of the International Union of Microbiological Societies (IUMS) and an interdisciplinary Commission of the International Union of Biological Sciences (IUBS).

The principal objective of the WFCC is to establish effective liaison between persons and organisations concerned with culture collections and the users of the collections both in the developed and developing regions of the world. To achieve this objective a structure of committees has been set up covering patents, postal and quarantine regulations, education, endangered collections, publicity and standards.

Committee on Patent Procedures. The activities of the Committee on Patent Procedures have important implications for biotechnology. The procedures for patenting processes involving the use of microorganisms, animal or plant cells or of genetically manipulated organisms are described in Chapter 6. The various patent regulations existing in different parts of the world present a confusing picture to those wishing to take out patents, and professional guidance is essential. A number of organisations such as the World Intellectual Property Organisation (WIPO) are concerned with the rationalisation of the different systems, and the WFCC's patents committee has acted in an advisory capacity to them, providing microbiological input. Members of the Committee have attended WIPO meetings to advise on the implementation of the Budapest Treaty for the International Recognition of the Deposit of Microorganisms for the Purpose of Patent Procedures. Additionally, they have monitored the functioning of the Treaty and provided evidence of difficulties that have arisen in its implementation. Recently a Guide to the Budapest Treaty has been published.

Committee for Quarantine and Postal Regulations. The Committee for Quarantine and Postal Regulations is similarly in close communication with the relevant postal regulatory bodies, such as the International Postal Union, National Postal Departments and the International Air Transport Association (IATA), and has put forward recommendations for the safe transport of infectious and non-infectious biological material (see Chapter 3). Members of the committee have been able to encourage international collaboration in this area by attending appropriate meetings and providing specialist advice in order to establish mechanisms for the safe transport of biological material throughout the world.

Education Committee. The WFCC is aware of the lack of guidance given to students before finishing university training on the support and services available from the microbial resource centres of the world. A similar lack of general awareness exists among many working microbiologists in industry, research and education. Accordingly, the Education Committee of the WFCC has an on-going programme of activities to increase the amount of information on back-up available from culture collections. Projects include the publication of books, preparation of training videos, advisory leaflets, and the organisation of training courses, scientific symposia and international conferences. This present series of source books is part of the programme of the Education Committee, designed to increase the usefulness of culture collections to those working in biotechnology.

A recent development has been the establishment of training schemes for individual scientists about to undertake extra responsibility in culture collections. Funds have been obtained from UNESCO and the International Union of Microbiological Societies to establish the first such schemes. Selected individuals have been given the opportunity to visit established collections to study administration and policy, 'shadowing' senior staff members and comparing different procedures and systems. It is expected that the scheme will lead to a greater awareness in young curators of the possibilities that exist in the establishment of new services and extended research initiatives.

Committee for Endangered Collections. The Committee for Endangered Collections is concerned to protect the microbial and cellular genetic resources of the world. Many of the major culture collections suffer from time to time from financial restrictions or from a change of direction in the interests of the host institute. Smaller collections are often transitory in nature and face difficulties on the retirement or relocation of the curator whose special interest the collection represents. The WFCC believes the conservation of these collections is of prime importance if the cultures and the substantial investment in terms of effort and expertise are not to be irretrievably lost. To enable emergency measures to be taken when difficulties arise, the Committee for Endangered Collections has obtained financial backing to assist in the provision of specialist, short-term support to allow the relocation of such collections to alternative laboratories willing and competent to take them over. The services of this committee may be used to provide advice to microbiologists who have developed collections of unique microorganisms during the course of their work, but who may not have the wish or expertise to maintain them in the long term or the resources to supply cultures to others.

Publicity Committee. The WFCC's Publicity Committee plays a major role in the dissemination of information about the activities of the Federation to the microbiological community. It produces a newsletter at regular intervals and is closely involved with all administrative developments. In particular, it plays an important part in the four-yearly WFCC International Conference and in the preparation of posters for scientific conferences. The editor of the newsletter will consider the publication of appropriate material and welcomes information about meetings, publications and topics of general interest to members. Biotechnologists may use the newsletter as a forum for the discussion of issues – possibly controversial – that are of interest to fellow scientists. Typical of subjects that can usefully be discussed in the columns of the newsletter are questions relating to stable nomenclature of microorganisms, the retention of published strain designations, security measures for the release of potentially dangerous cultures to those unqualified to handle them or the rescue of important genetic resources.

Committee for Standards. The WFCC is conscious of the fact that no authoritative standards exist for culture collections. Accordingly, a committee has been set up to prepare standards which can serve as guidelines.

It is recognised that it is not possible for collections from different parts of the world to reach the same standards, and there is no intention to impose standards or categorise collections according to their facilities. It is nevertheless felt that the existence of WFCC Guidelines on Standards for Culture Collections will serve as a valuable reference index and a stimulus to attain the highest standards possible within the economic limits of any collection. Very high professional standards may be reached with modest resources.

Data centres. In addition to the functions of these Committees, and others set up from time to time as the need arises, the WFCC has sponsored and is responsible for the World Data Center on Collections of Cultures of Microorganisms. The Center has pioneered the collection of data of this kind and has been responsible for the publication of three Directories listing the collections and the species they hold. The Center was originally housed in the University of Queensland's Department of Microbiology in Australia, but in 1986, on the retirement of its founder Director, was transferred to the Life Sciences Division at RIKEN, Tokyo, Japan. The WFCC is also co-sponsor with CODATA and IUMS of the international Microbial Strain Data Network (MSDN) set up to provide a referral system to the numerous data centres developing throughout the world listing microbial strain data. These two important activities of the WFCC, set up with international funding, are further discussed in Chapter 2.

The WFCC plays a prime role in the organisation of culture collection activities internationally and has among its membership experts in many areas of microbiology. It exists to serve both the culture collections and their users and may be used as a powerful interdisciplinary organ of communication between biotechnologists and specialists in other areas of microbiology.

8.2.2 *The MIRCEN Network*

The *UNESCO Courier* of July 1975 carried a feature 'On the road to development – a UNESCO network for applied microbiology'. Therein several mechanisms – conferences, training courses and fellowships – were identified. Since then, as a means towards strengthening the world network, several regional and international initiatives have been built in through the establishment of microbiological resources centres (MIRCENs) (see Table 8.1). These are designed:

(1) to provide the infrastructure for the building of a world network which would incorporate regional and interregional functional units geared to the management, distribution and utilisation of the microbial gene pool;
(2) to strengthen efforts relating to the conservation of microorganisms with emphasis on *Rhizobium* gene pools in developing countries with an agrarian base;
(3) to foster the development of new inexpensive technologies that are native to the region;
(4) to promote the applications of microbiology in the strengthening of rural economies;
(5) to serve as focal centres for the training of manpower and the imparting of microbiological knowledge.

The first development in the UNESCO global network of Microbiological Resource Centres, consisting of centres in the developed world

Table 8.1. *Microbial resource centres*

Biotechnology MIRCENs

Ain Shams University, Faculty of Agriculture, Shobra-Khaima, Cairo, Arab Republic of Egypt

Applied Research Division, Central American Research Institute for Industry (ICAITI), Ave La Reforma 4–47, Zone 10, Apdo Postal 1552, Guatemala City, Guatemala.

Bioconversion Technology MIRCEN, Caribbean Industrial Research Institute, Tunapuna, Trinidad and Tobago

Bioengineering MIRCEN, Centre de Transfer de Microbiologie Biotechnologie, UPS-INSA, Avenue de Rangueil, F–31077 Toulouse Cédex, France

Biotechnology MIRCEN, Department of Microbiology, University of Queensland, St Lucia, Queensland 4067, Australia

CAB International Mycological Institute, Mycology MIRCEN, Ferry Lane, Kew, Surrey TW9 3AF, UK

Department of Bacteriology, Karolinska Institutet, Fack, S–10401 Stockholm, Sweden

Fermentation, Food and Waste Recycling MIRCEN, Thailand Institute of Scientific and Technoloigcal Research, 196 Phahonyothin Road, Bangken, Bangkok 9, Thailand

Fermentation, Food and Waste Recycling MIRCEN, ICME, University of Osaka, Suita-shi 656, Osaka, Japan

Institute for Biotechnological Studies, Research and Development Centre, University of Kent, Canterbury CT2 7TD, UK

Marine Biotechnology MIRCEN, Department of Microbiology, University of Maryland, College Park Campus, Maryland 207742, USA

Microbial Technology MIRCEN, Institute of Microbiology, Academia Sinica, Zhongguanoun, Beijing 100080, China

Planta Piloto de Procesos Industriales Microbiologicos (PROIMI), Avenida Belgrano y Pasaje Caseros, 4000 S.M. de Tucuman, Argentina

University of Waterloo, Ontario, N2LK 3GI, Canada and University of Guelph, Guelph, Ontario NIG 2WI, Canada

Rhizobium MIRCENs

Cell Culture and Nitrogen-Fixation Laboratory, Room 116, Building 011–A, Barc-West, Beltsville, Maryland 20705, USA

Centre National de Recherches Agronomiques, d'Institut Sénégalais de Recherches Agricoles, BP 51, Bambey, Senegal

Departments of Soil Sciences and Botany, University of Nairobi, PO Box 30197, Nairobi, Kenya

IPAGRO, Postal 776, 90000 Porto Alegre, Rio Grande do Sul, Brazil

NIFTAL Project, College of Tropical Agriculture and Human Resources, University of Hawaii, PO Box 'O', Paia, Hawaii 96779, USA

World Data Center MIRCEN

World Data Center on Collections and Microorganisms, RIKEN, 2–1 Hirosawa, Wako, Saitama 351–01, Japan and regional networks in the developing countries, was the establishment of the World Data Center (WDC) on Microorganisms (see above and Chapter 2) in Queensland, Australia.

The MIRCEN at the Karolinska Institute, Sweden, in addition to developing microbiological techniques for the identification of microorganisms at the WDC, has pioneered the organisation of a series of MIRCENET Computer Conferences on biogas production, anaerobic digestion and the bioconversion of lignocellulose. Computer conferencing is a network system that links geographically scattered nodes together through the use of home or office computers to a remote control computer (see Chapter 2).

Apart from attempting to link up the MIRCENs and organising specialised conferences, MIRCENET has other functions listed in Table 8.2.

On the basis of their research and training programmes, the other MIRCENs can be broadly classified as follows.

The Biotechnology MIRCENs. In the area of biotechnology, there are 14 MIRCENs in operation (see Table 8.1). These are in Thailand, Egypt, Guatemala, Japan, Argentina, USA, the UK, Canada, Sweden, France, Australia, China, Trinidad and Tobago.

In the region of Southeast Asia, the MIRCEN in Bangkok has cooperating laboratories in the Philippines, Indonesia, Singapore, Malaysia and Hong Kong and other institutions in Thailand. It serves the microbiological community in the collection, preservation, identification and distribution of microbial germplasm, and in the promotion of research and training activities directed towards the needs of the region.

In the region of the Arab States, the MIRCEN at Ain-Shams University, Cairo, promotes research and training courses on the conservation of microbial cultures and biotechnologies of interest to the region.

Table 8.2. *Functions of MIRCENET*

To help initiate closed computer conferences under defined keys such as microbiology, biological nitrogen fixation, biogas, networking in culture collections.
To act as an information source for meetings, reviews, identification services, etc.
To provide a platform for discussions of MIRCEN network activities.
To provide print-outs and records of MIRCENET entries.

Through its cooperating MIRCEN laboratory at the University of Khartoum, the MIRCEN has contributed to the establishment of a culture collection in Sudan specialising in fungal taxonomy. The cooperating MIRCEN laboratory at the Institute Agronomique et Veterinaire Hassan II, Rabat, has made commendable progress through projects using different species of yeasts and rhizobia.

In the region of Central America and the Caribbean, the MIRCEN (cooperating laboratories in Chile, Columbia, Costa Rica, Dominican Republic, Ecuador, El Salvador, Honduras, Jamaica, Mexico, Nicaragua, Peru, Venezuela) has, in cooperation with the Organization of American States, the Interamerican Development Bank and several other prestigious agencies, pioneered the applications of microbiology, process engineering and fermentation technology in several member states of Central America and the Caribbean. It has set up joint collaborative research projects, the exchange of technical personnel, regional training programmes and the dissemination of scientific information among network institutions.

The South American Biotechnology MIRCEN located at Tucuman, Argentina comprises a regional network with cooperating laboratories in Brazil, Chile, Bolivia and Peru. It has similar goals to the biotechnology MIRCEN for Central America and the Caribbean.

The MIRCENs in the industrialised societies function as a bridge with those in the developing countries. In such a manner, increased cooperation is promoted between the developed and developing countries. Furthermore, a basic structure is set up for eventual twinning at a later date. For example, the Guelph Waterloo MIRCEN, Canada, with its expertise at the University of Waterloo in biomass conversion technology, microbial biomass protein production and bioreactor design, is of immense benefit to the work of the MIRCENs at Cairo, Guatemala and Tucuman.

In a similar manner, the MIRCEN at Bangkok has several collaborative research projects with that at the International Centre of Co-operative Research in Biotechnology, Osaka, Japan. This centre conducts the annual UNESCO International Postgraduate University Course on Microbiology (of 12 months' duration). It also functions as the Japanese point-of-contact for the Southeast Asian regional network of microbiology in the UNESCO Programme for Regional Co-operation in the basic sciences.

In the UK there is a MIRCEN network centred upon the Institute for Biotechnological Studies (IBS). In common with other Microbiological

Resource Centres, the aims of the UK MIRCEN are to promote the utilisation of the microbial gene pool, to promote applied microbiology and biotechnology in the developing countries and to provide a centre for training and advice.

The intergovernmental CAB International Mycological Institute (CMI) is the MIRCEN for mycology cooperating with all others in the field world-wide. This organisation, together with the Institute of Horticultural Research (IHR) are the first two organisations to collaborate with the MIRCEN Network, whilst continuing their own activities in mycological and biodeterioration studies, and microbiological pest control, mycorrhizas and mushroom technology respectively.

In keeping with the new trends of the expanding frontiers of biotechnological research, a Marine Biotechnology MIRCEN has been established at the University of Maryland. Work presently under way includes the fundamental elucidation of the evolution of genes and the flow of genes through populations in the marine environment. One collaborative study under way is with the Chinese University of Hong Kong and the Shandong College of Oceanography, Qingdao, China.

The Biological Nitrogen Fixation (BNF) MIRCENS. In the quest for more food for their increasing populations, several developing nations have been expanding their agricultural lands into areas which are marginally capable of sustaining productivity and invariably limited by the availability of nitrogen fertilizer.

In interaction with other international programmes, modest schemes for the development of biofertilizers or *Rhizobium* inoculant material, particularly in legume-crop areas of the developing countries, are already operating through the MIRCENs on a level of regional cooperation in Latin America, East Africa and Southeast Asia and the Pacific.

In the area of biological nitrogen five MIRCENs are already operating (see Table 8.1). The broad responsibilities of these MIRCENs include collection, identification, maintenance, testing and distribution of rhizobial cultures compatible with crops of the regions. Deployment of local rhizobia inoculant technology and promotion of research are other activities. Advice and guidance are provided in the region to individuals and institutions engaged in rhizobiology research.

The BNF MIRCENs play a valuable role in maintaining and distributing efficient cultures of *Rhizobium*. Nearly 4000 strains are maintained in the MIRCEN collections and about 1750 have been distributed to other organisations (Table 8.3).

The MIRCEN network is founded on the principle of self-help and mutual cooperation. It concentrates on existing facilities and resources and provides an organisational structure which allows each institution to collaborate as best it can through the following:

(1) an exchange of research workers between national and regional institutions;
(2) small grants to individual research projects or workers for acquisitions of supplies, spare parts for equipment, or small-scale equipment;
(3) participation of senior scientists in specialised symposia in

(*contd. on p. 176*)

Table 8.3. *Culture collection services of Biological Nitrogen Fixation (BNF) MIRCENs*[a]

Holdings of Rhizobium *culture collections*

MIRCEN	Number of strains held
Bambey	50
Beltsville	938
Hawaii	2000
Nairobi	208
Porto Alegre	650
Total	3846

Cultures distributed by Rhizobium *MIRCENs*

MIRCEN	Number of cultures	Countries of recipient institutions
Bambey	8	Gambia, Mali, Yemen
Beltsville	508	Zimbabwe, Nigeria, Yugoslavia, India, Spain, Vietnam, Ireland, UK, Malaysia, Italy, Canada, South Africa, Senegal, Egypt, Poland, Argentina, Turkey, W. Germany, Austria, Australia, New Zealand
Hawaii	200	Global
Nairobi	95	Uganda, Malawi, Tanzania, Mauritius, Sudan, Congo, Zaïre, Rwanda
Porto Alegre	943	Argentina, Chile, Bolivia, Uruguay, Peru, Ecuador, Colombia, Venezuela, El Salvador, Dominican Rep., Mexico, USA, Trinidad, Brazil

[a] 1988 figures.

the technically advanced countries in the vicinity of each of the regions;

(4) organisation of short-term intensive training courses and specialised in-depth sub-national or national meetings;

(5) production of a newsletter functioning as an outlet for the exchange of research news, publication of research findings and as an attraction for potential participating laboratories.

The MIRCENs play a catalytic role in breaching the barrier of geographical isolation and advancing the frontiers of contemporary research in biotechnology through the production of newsletter bulletins, culture collection catalogues and research papers. The publication of *MIRCEN News* annually, the development of MIRCENET, and the UNESCO *MIRCEN Journal of Applied Microbiology and Biotechnology* (now incorporated in the *World Journal of Microbiology and Biotechnology*) are indications of the gradual emergence of competence and capability of the MIRCENs, and the services they provide on a regional and interregional basis.

8.3 Regional organisation

Transnational coordinating mechanisms are being set up throughout the world to bring regional cohesion to culture collection activities and benefit to both the resource centres themselves and their users. Some have been established as committees by culture collections; others have originated as data centres with the secondary effect of stimulating closer working collaboration between the contributing culture collections. They may be contacted for information about microbiological resources, services and general advice.

8.3.1 *European Culture Collections' Organisation (ECCO)*

In 1981, at an international conference in Brno, Czechoslovakia, curators of European service culture collections present agreed that a mechanism should be set up to enable meetings to take place on an annual basis for the exchange of ideas and the discussion of common problems. In 1982 the first meeting of ECCO took place at the Deutsche Sammlung von Mikroorganismen, Göttingen, Federal Republic of Germany, and since then meetings have been held in France, the UK, Czechoslovakia, Hungary, the Netherlands and the USSR. Membership has increased steadily as new culture collections are formed or developed to provide a national service. Membership is restricted to collections that provide a service on demand and without restriction, that have as a normal part of their duty the acceptance of cultures, that issue from time to time a list of their holdings, and that are in a country with a microbiological society belonging to the Federation of European Microbiological Societies. It was felt that these collections have interests and problems in common that are not shared by research or teaching collections. The Organisation is affiliated to FEMS and an adherent member of the World Federation for Culture Collections.

Apart from the exchange of scientific information relating to such topics as taxonomy, identification and preservation procedures, much benefit has been derived from discussions on new developments within culture collections such as acceptance of International Depositary Authority status (see Chapter 6) or the development of computerised systems for the storage, searching and dissemination of culture information (see Chapter 2). In addition, the opportunity to meet on a regular basis has enabled collaborative programmes to be set up between collections from different countries. Discussions are taking place on the possible expansion of ECCO to increase scientific activities and broaden representation.

ECCO members have become aware that the services available from the collections are not fully exploited by users. To remedy this they have combined to produce publicity material which may be obtained from the recently established Information Centre for European Culture Collections (ICECC) at Braunschweig, Federal Republic of Germany (see Chapter 2). Information about the holdings and services of ECCO collections is also available from the Organisation's Officers (see Chapter 2) or the Secretary of FEMS.

8.3.2 *Regional database systems*

A number of coordinating mechanisms based on information centres have been set up, primarily to establish data banks for regional access. These are described in greater detail in Chapter 2.

Some, such as the Tropical Data Base in Brazil, the Microbial Information Network Europe (MINE) and the Nordic Register, have been developed initially to provide a centre for information on the culture collections themselves, their services and their holdings. Others, such as the Microbial Culture Information Service (MiCIS), have been set up in areas with well established culture collection systems with the purpose of providing on-line strain databases for searching. Both these kinds of data centres have the secondary effect of encouraging collaboration between culture collections so that the best possible system develops and minimum duplication of effort takes place.

The proliferation of microbial data centres world-wide reflects the growing need for information on biological materials. It has also led to problems in identifying the most appropriate point of contact for specific information. To overcome this an international system has been established (Microbial Strain Data Network, MSDN), to act as a referral system and communications network to databases able to answer specific enquiries on strain properties (Chapter 2). It seems certain that other systems will be established, and the function of the MSDN thus becomes of increasing importance as the first point of enquiry, directing those seeking information on strain properties to appropriate centres.

8.4 National federations and committees

The following countries have established federations or committees for the coordination of culture collection activities:

Australia
Canada
China
Czechoslovakia
Japan
Korea
New Zealand
Turkey
United Kingdom
United States of America

Information about them and their activities may be obtained through culture collections or microbiological societies within the country or through the World Data Center and the Microbial Strain Data Network (see Chapter 2). Most of these organisations produce newsletters from time to time and further information may be obtained through these publications.

Some of the organisations are *for* culture collections, others are *of* culture collections and the difference between the two categories is significant. Those that are *of* culture collections exist primarily to coordinate culture collection activities within the countries (produce common catalogues, rationalise holdings, stabilise funding) and are generally termed Committees rather than Federations; those that are *for* culture collections have as their prime function the promotion of communication between the collections and their users in industry, research and education. The activities of the latter category concentrate more on scientific meetings, workshops and training courses, and the membership includes any microbiologists with an interest in culture collection activities, whether they are working in culture collections or not. The executive boards are deliberately formed of people both from culture collections and from research or teaching laboratories and industry, providing a cross fertilisation of interests, whereas with organisations set up *for* culture collections the officers and members are drawn from the collections only. The impact of biotechnological input to the Federations has played a valuable part in the development of resource centres to meet the growing needs of industry in this area.

A number of international organisations exist for the coordination of activities within different microbiological disciplines, and information about them can be obtained from the International Council for Scientific Unions (Table 8.4). Information on biotechnology is disseminated through the different associations listed in Table 8.5. All these organisations recognise the need for an effective network of microbial resource centres and are active in support of their development.

8.5 Future developments

Developments in biotechnology have coincided with extensive advances in computer technology, and throughout the world culture collections have taken advantage of the latter to respond to the increasing demands of the former. It is clear from Chapter 2 that data held in the microbial resource centres is increasingly computerised and it is

Table 8.4. *International scientific organisations*

ICSU	International Council of Scientific Unions 51 Boulevard de Montmorency F–75016 Paris France Telephone: (1) 45-25-03-29 Telex: ICSU 630553 F Electronic mail: TELECOM GOLD 75:DBI0126
IUBS	International Union of Biological Sciences
IUMS	International Union of Microbiological Societies
ICRO	(International Cell Research Organisation) Panel on Applied Microbiology and Biotechnology

evident that the biotechnology community can better be served by coordination of these activities. The World Data Center for Collections of Cultures of Microorganisms and the Microbial Strain Data Network are important examples of international collaboration in this area, leading to on-line databases and information network systems. The imaginative and successful MIRCEN network will continue to be instrumental in encouraging the establishment and development of culture collection activities in the developing world and linking them to those in industrial nations.

Table 8.5. *Biotechnology associations*

ABA	Australian Biotechnical Association 1 Lorraine Street Hampton Victoria 3188 Australia
ABC	Association of Biotechnology Companies 1220 L Street NW Suite 615 Washington, DC 20005 USA
ADEBIO	Association de Biotechnologie 3 rue Massenet F–77300 Fontainebleau France
BIA	BioIndustry Association 1 Queen's Gate London SW1H 9BT UK
BIDEC	C/o Japan Association of Industrial Fermentation 20–5 Shinbashi 5-chome Minato-ku Tokyo 105 Japan
IBA	Industrial Biotechnology Association 2115 East Jefferson Street Rockville Maryland 20852 USA
IBAC	Industrial Biotechnology Association of Canada Lava University Cité Universitaire Québec G1K 7P4 Canada

Computers will be used increasingly for computer conferencing and electronic mail, leading to greater communication between the collections. This in turn should lead to greater collaborative research and joint service activities and will minimise unnecessary duplication of effort, consistant with national requirements.

In spite of rapid developments in communication systems, the need for the presence of culture collections in all regions of the world will remain because of specialised local needs, regional regulatory requirements, such as those for postal and quarantine purposes, or currency or language reasons. Duplication of important holdings and services is necessary, but can be reduced to an acceptable level by collaborative efforts on the part of individual scientists in the resource centres, the setting up of organisations to coordinate their activities and the use of computers and electronic networking to facilitate communication. A basic core of collaborative mechanisms already exists and can be extended to cover regions of the world or specialist areas of activity not yet coordinated internationally.

INDEX